AF714344

Climate Change Governance and Adaptation

Climate Change Governance and Adaptation

Case Studies from South Asia

Edited by

Anamika Barua

Vishal Narain

Sumit Vij

CRC Press
Taylor & Francis Group
Boca Raton London New York

CRC Press is an imprint of the
Taylor & Francis Group, an **informa** business

CRC Press
Taylor & Francis Group
6000 Broken Sound Parkway NW, Suite 300
Boca Raton, FL 33487-2742

First issued in paperback 2020

© 2019 by Taylor & Francis Group, LLC
CRC Press is an imprint of Taylor & Francis Group, an Informa business

No claim to original U.S. Government works

ISBN 13: 978-0-367-57112-2 (pbk)
ISBN 13: 978-1-138-05450-9 (hbk)

This book contains information obtained from authentic and highly regarded sources. Reasonable efforts have been made to publish reliable data and information, but the author and publisher cannot assume responsibility for the validity of all materials or the consequences of their use. The authors and publishers have attempted to trace the copyright holders of all material reproduced in this publication and apologize to copyright holders if permission to publish in this form has not been obtained. If any copyright material has not been acknowledged please write and let us know so we may rectify in any future reprint.

Except as permitted under U.S. Copyright Law, no part of this book may be reprinted, reproduced, transmitted, or utilized in any form by any electronic, mechanical, or other means, now known or hereafter invented, including photocopying, microfilming, and recording, or in any information storage or retrieval system, without written permission from the publishers.

For permission to photocopy or use material electronically from this work, please access www.copyright.com (http://www.copyright.com/) or contact the Copyright Clearance Center, Inc. (CCC), 222 Rosewood Drive, Danvers, MA 01923, 978-750-8400. CCC is a not-for-profit organization that provides licenses and registration for a variety of users. For organizations that have been granted a photocopy license by the CCC, a separate system of payment has been arranged.

Trademark Notice: Product or corporate names may be trademarks or registered trademarks, and are used only for identification and explanation without intent to infringe.

Library of Congress Cataloging-in-Publication Data

Names: Barua, Anamika. | Narain, Vishal. | Vij, Sumit.
Title: Climate change governance and adaptation : case studies from South Asia / [edited by] Anamika Barua, Vishal Narain, and Sumit Vij.
Description: Boca Raton : CRC Press, Taylor & Francis Group, 2018. | Includes bibliographical references.
Identifiers: LCCN 2018014590 | ISBN 9781138054509 (hardback : alk. paper)
Subjects: LCSH: Climatic changes--South Asia. | Climatic changes--India. | Climatic changes--Social aspects--South Asia. | Climatic changes--Social aspects--India. | Conservation of natural resources--South Asia. | Conservation of natural resources--India.
Classification: LCC QC903.2.S64 C55 2018 | DDC 363.738/745610954--dc23
LC record available at https://lccn.loc.gov/2018014590

Visit the Taylor & Francis Web site at
http://www.taylorandfrancis.com

and the CRC Press Web site at
http://www.crcpress.com

"To the communities of South Asia"

Contents

Foreword ... ix
Acknowledgments ... xi
Editors ... xiii
Contributors ... xv

1. **Governance of Climate Change: Issues and Challenges in South Asia** ... 1
   Anamika Barua, Vishal Narain, and Sumit Vij

2. **"A Disappearing" or a Resilient Ganga? Climate Change Perspectives from the Himalaya** ... 11
   Georgina Drew

3. **Deliberative Governance on Vulnerability to Climate Change: Voices from Madhesi Farmers** ... 29
   Floriane Clement

4. **Waste or Savior? Two Cases of Emerging Wastewater Irrigation in Urbanizing Kathmandu Valley** ... 51
   Silvia Quarta, Dik Roth, Robert Dongol, Anushiya Shrestha, and Saroj Yakami

5. **Climate Migration and Flood Related Disasters** ... 71
   Anna Wesselink, Jeroen Warner, Michelle Kooy, Parvin Sultana, Paul Thompson, and Abu Syed

6. **Changing Social Capital in the Mountains and the Implications for Adaptation Interventions: An Exploratory Analysis with Case Studies from the Hindu Kush Himalaya** ... 89
   Arabinda Mishra

7. **Enhancing Adaptive Capacity through Education: A Case Study of Rural Mountain Communities, Sikkim, Eastern Himalaya India** ... 109
   Bhupen Mili, Suparana Katyaini, and Anamika Barua

8. **Governing National Actions for Global Climate Change Stabilization: Examples from India** 137
Joyashree Roy, Shyamasree Dasgupta, Duke Ghosh, Nandini Das, Debalina Chakravarty, Debrupa Chakraborty, and Sudipta De

9. **Bridges and Barriers for Governance of Water Resources in South Asia—Insights from Brahmaputra Basin in India** 161
Navarun Varma and Sanjoy Hazarika

Index 183

Foreword

This age is known as the Anthropocene, where an unprecedented scale of human activity has profusely impacted environmental change and caused enormous stress on natural resources. There is now established evidence of climate change; this evidence is seen in an increased incidence of extreme events, changes in the volume and intensity of rain and snowfall, changes in the seasonal distribution of rainfall, and a shift in the duration of seasons. Donors and funders have invested huge sums of financial resources to build the capacity of individuals to adapt to climate change and the capacity of governments and policy-makers to deal with the challenges caused by it. Disaster risk reduction, vulnerability, adaptation, and resilience are the buzzwords in the discourse on climate change, engaging the attention of researchers, students, academics, and development practitioners.

There is a need for a greater and deeper understanding of the relationships between governance and climate change. When we say governance, we refer to the relationships among different actors—the state, markets, and civil society—and do not stay within the narrow confines of the government. In this manner, this book explores a set of issues around governance and climate change in South Asia. How is climate change redefining the roles of different actors in governance processes? How are relationships of control and authority over natural resources changing? How are social relationships and networks being transformed? What factors shape how communities respond to climate change? What is the role of human perceptions and how can the plans for climate change adaptation better incorporate community perceptions and perspectives? Contributions in this book address these questions and a few other ones.

Several things make this book unique. First, although the regional focus is on South Asia, it draws on a wide range of contributors from northern and southern researchers cutting across various social science disciplines. Second, contributors cut across both the academic and practitioner communities and this interface is necessary to find solutions to the problems caused by climate change. This outreach allows an articulation of different points of view. Third, contributions are from different agro-ecological and institutional contexts. This difference allows coverage of a diversity of perspectives and issues such as the transformations in social capital brought about by climate change in the Hindukush Himalaya, the role of deliberative governance in climate change adaptation, institutional innovation in responding to water insecurity induced by urbanization and climate change in peri-urban spaces, and the (potential) role of human perceptions of climate change in shaping the framing and effectiveness of climate change adaptation plans.

Given the diversity of themes covered, this book will be of interest to a wide cross-section of researchers. Perhaps the greatest value addition of the book will be in defining a set of themes that can constitute the frontiers of a research program on climate change adaptation in South Asia. This book will appeal not only to researchers and practitioners in the climate change community, but also to those with a broader interest in governance processes. By keeping a focus on governance and not on government, the book does not seek to absolve state institutions of the responsibility for climate action; on the other hand, an understanding of autonomous adaptation responses can be used to inform planned adaptation responses. An understanding of how societal institutions such as social capital are transformed in response to climate change can reveal deeper insights into the relationship between ecological/biophysical and institutional processes and can be instrumental in drawing further attention to research to unravel these links that are multi-causal and rarely linear.

To conclude, this is a timely addition to the literature on climate change adaptation in general and in South Asia in particular. What would be interesting to see is how it sets the base for future climate change adaptation research in the region, urging the contributors in the volume—as well as outside it—to refine some of the themes highlighted in the book, but also to use it as a basis for engaging with policy-makers and development practitioners in the region.

Saleemul Huq
International Centre for Climate Change and Development (ICCCAD), Bangladesh
International Institute for Environment and Development (IIED), London

Acknowledgments

South Asia is vulnerable to the impacts of climate change; hence, responsive and inclusive climate governance is crucial. This book on climate governance in South Asia brings in the collection of researched cases from South Asia reflecting the need to understand the relationship between governance and climate change at multiple levels.

We would like to thank each of the contributors to this book for sharing their experience and knowledge through their chapters in this book. Their view points on governance which is reflected in their contribution play an important role in bridging some of the gaps in understanding climate governance. We sincerely and gratefully acknowledge the time and the effort they have devoted to this book.

We are grateful to Dr. Saleemul Huq for agreeing to write a foreword to this book and for acknowledging the need for such a volume, which can set the ground for future climate governance research in South Asia.

The editors would also like to thank the editorial team at CRC Press for publishing this book and for their support throughout the project.

Editors

Anamika Barua is an associate professor at the Department of Humanities and Social Sciences, IIT Guwahati, India. She has a PhD in ecological economics from the University of Leeds, UK. Her work focuses on understanding how political, social, and economic factors shape environmental decisions and change, particularly related to water. For the last 10 years she has been involved in academic and consultancy project related to water issues in South Asia. Her research interests include—economics of climate change, ecological footprints, virtual water flow through trade, and water governance including transboundary water governance. She has published her recent research in journals such as *Regional Environmental Change, Water Resources Research,* and *International Journal for Water Resources Development and Water Policy.*

Vishal Narain is a professor at Public Policy and Governance Department, Management Development Institute, India. He holds a PhD from Wageningen University, the Netherlands. His teaching and research interests are in the inter-disciplinary analyses of public policy processes and institutions, water governance, vulnerability and adaptation to environmental change, and peri-urban issues. His research has been published in several international peer-reviewed journals such as *Water Policy, Water International, Environment and Urbanization, Local Environment: the International Journal of Justice and Sustainability,* and *Mountain Research and Development.* He has recently co-edited "Globalization of Water Governance in South Asia (Routledge)." He was the lead author for a chapter on human vulnerability to environmental change for Global Environment Outlook-4 (GEO-4), the flagship publication of UNEP, the United Nations Environment Programme.

Sumit Vij is a PhD researcher at Public Administration and Policy Group in Wageningen University, the Netherlands. His research interests are climate governance, power and politics, policy paradigms, transboundary water governance and politics, and peri-urban development. Sumit holds a master's degree in Sustainable Development Practices from TERI University, New Delhi, India, and studied Development Management at TDA. He completed a bachelor's degree in Chemistry from University of Delhi, India. His experience also includes his service record at the National Dairy Development Board, Ministry of Agriculture, India. He has recently published articles in journals such as *Environment Science and Policy, Climate Policy, International Journal for Water Resources Development, Land Use Policy,* and *Geoforum.*

Contributors

Debrupa Chakraborty
Department of Commerce
Netaji Nagar Day College
Kolkata, India

Debalina Chakravarty
Department of Economics
National Institute of Technology, Jamshedpur
Jamshedpur, India

Floriane Clement
INRA, UMR Dynafor
Auzeville CS, France

Nandini Das
Department of Economics
Jadavpur University
Kolkata, India

Shyamasree Dasgupta
Department of Humanities and Social Sciences
Indian Institute of Technology Mandi
Mandi, India

Sudipta De
Department of Mechanical Engineering
Jadavpur University
Kolkata, India

Robert Dongol
Department of Civil Engineering
Nepal Engineering College
Bhaktapur, Nepal

Georgina Drew
Anthropology and Development Studies
University of Adelaide
Adelaide, Australia

Duke Ghosh
Global Change Programme
Global Change Research
Kolkata, India

Sanjoy Hazarika
Commonwealth Human Rights Initiative
New Delhi, India

Suparana Katyaini
Department of Humanities and Social Sciences
Indian Institute of Technology Delhi
New Delhi, India

Michelle Kooy
Water Governance
UNESCO-IHE
Delft, the Netherlands

Bhupen Mili
Department of Humanities and Social Sciences
Indian Institute of Technology Guwahati
Guwahati, India

Arabinda Mishra
International Centre for Integrated Mountain Development
Patan, Nepal

Silvia Quarta
Department of Earth System and Science
Wageningen University
Wageningen, the Netherlands

Dik Roth
Sociology of Development and Change
Wageningen University
Wageningen, the Netherlands

Joyashree Roy
Department of Economics
Jadavpur University
Kolkata, India

Anushiya Shrestha
Sociology of Development and Change
Wageningen University
Wageningen, the Netherlands

Parvin Sultana
Flood Hazard Research Centre
Dhaka, Bangladesh

Abu Syed
Bangladesh Centre for Advanced Studies
Dhaka, Bangladesh

Paul Thompson
Flood Hazard Research Centre
Dhaka, Bangladesh

Navarun Varma
Residential College 4
National University of Singapore
Singapore

Jeroen Warner
Disaster Studies, Sociology of Development and Change
Wageningen University
Wageningen, the Netherlands

Anna Wesselink
Department of Integrated Water Systems and Governance
UNESCO-IHE
Delft, the Netherlands

Saroj Yakami
MetaMeta Nepal
Patan, Nepal

1

Governance of Climate Change: Issues and Challenges in South Asia

Anamika Barua, Vishal Narain, and Sumit Vij

CONTENTS

1.1 Introduction 1
1.2 Climate Change Governance and Societal Challenges in South Asia 3
1.3 Adaptation to Climate Change-Induced Impacts through Effective Governance 5
1.4 Climate Change Governance at the National Level and Regional Level 6
1.5 Conclusion 7
References 9

1.1 Introduction

Climate change and its impacts threaten the natural resource-based livelihoods of South Asia, putting poor and marginalized communities at risk (Adger et al. 2003; Huq et al. 2004). Prompt action is necessary to reduce the causes of climate change and adapt to its impacts. Interest in adaptation and mitigation as response strategies to the challenges posed by climate change is increasing. However, adaptation and mitigation are cross-sectoral in scope and thus demand responses from multiple levels of government (Gemmer et al. 2011). Therefore, a deeper understanding of how climate change processes affect governance processes at multiple levels is necessary, as well as how new institutions or existing institutions are adapting to the challenges posed by climate change. Further, an understanding of how communities perceive climate change and how they innovate—technologically or institutionally—to adapt to the impacts of climatic changes on their livelihoods is necessary as the basis of policy approaches to climate change adaptation. Similarly, effective mitigation policies must be coordinated at local, national, and regional levels.

The Global North has continuously debated climate change as an environmental and social problem. Recently, research in the Global South has emphasized issues of vulnerability and community-based adaptation, mostly at the local level. Issues such as governance of adaptation at different levels remain poorly studied in the context of South Asia (Vij et al. 2017). Much of the focus is on development strategies without emphasis on long-term characteristics. Moreover, technocratic, top-down, and aid-driven strategies dominate climate change governance without focusing on values or the societal relevance of climate change risk reduction options (Ojha et al. 2016). The lack of focus on socio-political relevance neglects the importance of climate change governance and policy research in the countries of South Asia.

With this understanding and background, this book, *Climate Change Governance: Perspectives and Experiences from South Asia*, addresses cases from South Asia on the governance of climate change adaptation and mitigation at the local, national, and regional levels. In this book, the concept of governance denotes the totality of interactions among actors involved in the steering of efforts at the regional, national, and local levels to deal with the impacts of climate change. Since the late 1980s, the paradigm shifted from a focus on government to a focus on governance (Mathur 2008). This focus on governance provides greater democracy and space to the states in policy making through involvement of local authorities, non-state actors such as civil society, parastatals, and international donor agencies, media, and communities (Ayling and Gunningham 2017; Kuyper et al. 2018).

Corresponding to this shift, this book focuses more broadly on governance to look at the role of actors beyond the state since the impacts of climate change spread well beyond the frontiers of the state; thus, communities and individuals respond in varying capacities to deal with these impacts. While examining the governance dimensions of climate change, this book attempts to examine the policy processes at the different levels. At the local level, for example, climate change interacts with rural-urban, rural-rural, and cross-country migration to impact livelihoods and gender relations, and communities and individuals are forging new policies and partnerships to adapt to and mitigate the impacts of climate change. At the national, this book examines how to align national policies with the global climate change mitigation goals and how to share policy responsibilities between different levels of governance.

While transboundary institutions (both formal and informal) can promote and enhance cooperation, they often lack capacity to include climate change uncertainty. The inability to include uncertainty may lead to disputes or conflicts among the riparian communities and can also impact the long-term stability of the existing treaties. This possibility is particularly true for South Asian rivers like the Indus (India and Pakistan) and Ganges (India and Bangladesh) where the countries sharing the respective rivers signed bilateral treaties to cooperate in managing the river but now upstream and downstream countries consider these treaties unfair due to uncertainty in the availability of water. Therefore, institutions need to evolve with time so that they have substantive adaptive capacity to mediate such issues (Wolf 2006).

This book aims to provide an understanding of the challenges and opportunities for increasing the resilience of the society through effective governance of climate change impacts. Further, this book emphasizes that a top-down approach to the governance of climate change adaptation and mitigation may not yield desired outcomes and instead emphasizes the need to integrate issues of equity, gender, the transformation of social capital, and the concerns of vulnerable groups into climate change governance and polices.

Each chapter includes a case study for a total of eight case studies that illustrate specific elements of climate change governance. This book classifies these chapters into three broad sections: (1) climate change governance and societal challenges in South Asia, (2) adaptation to climate change-induced impacts through effective governance, and (3) climate change governance at national and regional levels.

1.2 Climate Change Governance and Societal Challenges in South Asia

The first section of this book addresses the specific gaps and challenges for governance at multiple levels and emphasizes the need to meaningfully engage communities in governance. Apart from providing a background to the cases discussed on climate change governance in South Asia, the chapters in this section emphasize the need to ensure social equity and environmental justice and stress the need for acceptance of adaptation and mitigation measures at the local level to consider people's values and belief systems.

Chapter 2 by Georgina Drew focuses on the North Indian state of Uttarakhan and makes a case for a more nuanced understanding of people's perceptions of climate change. The author argues that climate change is a phenomenon that one needs to study with an interpretive lens, looking at the meaning that people attach to the process. It is a phenomenon that one should study beyond measurements of hydro-meteorological data. While glaciologists and climatologists focus on changing glacier melt and river flows, the meaning that people give to these processes shape the adaptive responses of communities. Using an ethnographic approach spread out over several years, the author shows the symbolic and religious association that people have with the Ganga river and how it shapes the meaning that they give to changing river flows. For example, Ganga, the eternal Goddess, can never disappear; instead, she goes into *Pataal* (underground). The chapter makes a case for a more locally based approach to climate change adaptation. Adaptation programs and policies that consider people's relationship with the resource (Ganges river in this case) as well as the meaning that the people give to the climate change phenomenon are likely to be better received and more sustainable. The larger argument is perhaps that climate change and its impacts should not be value neutral, positivist processes, but rather socially constructed processes in which people's values and belief systems are center stage.

This approach then brings us to an important question of how to operationalize this process. If climate change is a social construction, then how does this knowledge and understanding of people's perceptions of climate change translate into something that policy-makers and planners can use? Chapter 3 by Floriane Clement focuses on Nepal and describes an approach of deliberative governance. As part of an action research project, the team developed an initiative to capture farmers' meanings and experiences of vulnerability. The research tried to capture this process by describing the dominant narratives and discourses that shaped the communities' experiences of climate change. This chapter demonstrates that while consensus building is a difficult challenge in processes of deliberative governance, the process helps in developing a shared understanding of the multiple framings surrounding climate change. Despite the limited number of interactions with the Madhesi farmers that the team engaged in, processes of deliberative governance do help in building relationships among stakeholders. Policy approaches to climate change adaptation should be based on what the people need the authors argue, rather than on prescriptive, top-down underpinnings.

Chapter 4 by Silvia Quarta, Dik Roth, Robert Dongol, Anushiya Shrestha, and Saroj Yakami explores changing wastewater use in a context of climate change and urbanization. The chapter is set in peri-urban Kathmandu. Peri-urban communities face the dual impacts of both urbanization and climate change. Recent years have seen an increase in the eroding water access for peri-urban communities in the region (for a review, see Narain and Prakash 2016). Peri-urban areas are spaces in transition, revealing a mix of both urban and rural characteristics. Peri-urban communities bear the brunt of urban expansion, losing land and water resources in the process. Climatic change and variability further aggravate these threats on the livelihoods of peri-urban communities, for example, through changing precipitation patterns or the changing seasonal distribution of rainfall. Peri-urban spaces, therefore, provide very interesting contexts to study climate change governance and adaptation because these peri-urban communities suffer from what O'Brien and others (2002) call double exposure.

It is in this context that Chapter 4 explores the evolving practices of peri-urban farmers around Kathmandu. The chapter demonstrates how water users innovate institutionally and technologically to make use of the opportunities created by the flow of wastewater from the city. This chapter contributes to a deeper understanding of the practices associated with the use of wastewater even though there is a burgeoning literature around its productivity and livelihood impacts (see, for example, IWMI 2003, 2006). Wastewater studies are increasingly conducted with a very instrumental perspective. However, to understand its widespread use in the region, it is important to understand the institutions—norms, practices, and codes of conduct—surrounding the use of the resource that make it accessible to more users. Wastewater will continue to be an important resource for peri-urban communities. On the one hand, the urbanization process and the growth and expansion of cities in the future will continue to generate large amounts of wastes. On the other hand, the reliance of peri-urban communities on this resource will continue to grow as access to other irrigation sources decreases (such as decline in water tables because of pressures from competing uses, and loss of tubewells and other irrigation sources because of land acquisition to support urban expansion).

1.3 Adaptation to Climate Change-Induced Impacts through Effective Governance

The second section of the book elaborates on adaptation strategies in South Asia. Through the case studies in this section, contributors emphasize the response of communities to adapt to climate change in the region and the factors that can enhance the capacity of communities to adapt to climate change.

Chapter 5 by Anna Wesselink, Jeroen Warner, Michelle Kooy, Parvin Sultana, Paul Thompson, and Abu Syed discusses climate change migration as a positive outcome mediated through such factors as local governance structures, policy responses, and socio-economic aspects, rather than climate change. Situating the case of migration in contemporary Bangladesh, the authors analyze the general problems of international migration framings and compare these narratives to the numerous migration-focused research projects and policies of Bangladesh. This chapter reviews key studies focusing on land grabbing and displacement as the reasons for migration. They argue that the lack of research and explanations for the relationship between climate change and migration results in missing out on issues of power and politics at the community level. This chapter concludes that climate change strategies and policies should go beyond science-based designs or the usual top-down policy-making processes if the goal is to move forward in supporting equitable adaptation to climate change.

Chapter 6 by Arabinda Mishra addresses the importance of social capital for improving community-level adaptive capacity to reduce the impacts of disasters, disruptions, and resource scarcity. The aim of this chapter is to explain the influences acting on the sources of social capital in rural mountain communities in the Hindu Kush Himalaya (HKH) region using five case studies. The author engages with the proposition that multiple forces are interacting with each other to produce a process of simultaneous decay and emergence of social capital in the communities, and that this has implications for planned adaptation interventions. Based on five case studies, the author argues that erosion of social capital is certainly bad news for adaptation governance. The weakening of resource managing institutions results in a risk of conflict among users and the elite capture of the institutions. Moreover, as depletion of social capital continues in a community, the transaction costs of cooperation on adaptation actions increase. Reduced social capital also results in the exclusion of marginalized groups from resource systems. Social capital emerges by enabling communities to overcome the hardships of a crisis and recover in a better manner, especially in improving adaptation and resilience building. The author concludes by emphasizing that public authorities should consider transformations in social capital in their climate change adaptation efforts.

Chapter 7 by Bhupen Mili, Suparana Katyaini, and Anamika Barua emphasizes strengthening human capacity to combat the adverse impacts of climate change. Considering education as an important aspect of human capacity, this chapter seeks to explore the role of education in shaping adaptive capacity among the rural mountain communities in the Eastern Himalayan region of India. The authors conducted household surveys and Focus Group Discussions (FGDs) to capture the linkage in the east district of Sikkim. The assessment, carried out to develop the thematic index on education, revealed high deprivation in education. The paper addresses social and gender differentiation in educational achievements. Social classification and gender differentiate the educational achievements because of traditional socio-cultural norms. Therefore, the authors emphasize the importance of designing long-term interventions rather than ad-hoc interventions in education because social change is an essential and time-consuming process for building adaptive capacity. This chapter also emphasizes the need to integrate the heterogeneity of the society in the design of the intervention because it mediates the conversion of resources (for example, investment in education) into actual achievements (Robeyns 2013). Therefore, the authors argue for designing interventions in education with a larger human development perspective by removing the socio-cultural barriers and by looking beyond provision of resources towards building capabilities of the rural mountain communities to enhance their adaptive capacity.

1.4 Climate Change Governance at the National Level and Regional Level

Climate change impacts large populations and this impact necessitates access to a range of approaches to enable policy actors to allocate scarce resources to reduce the consequences of these impacts. The third section of this book, therefore, focuses specifically on various challenges to developing policies and programs that the global agenda usually drives.

Chapter 8 by Joyashree Roy, Shyamasree Dasgupta, Duke Ghosh, Nandini Das, Debalina Chakravarty, Debrupa Chakraborty, and Sudipta De focuses on the climate governance issue with examples from India using ideas from a multi-level governance framework and a socio-technical transition theory. They describe the vertical alignments and actors involved in the climate governance structure in India. The authors follow the nature and quality of articulation of actions and role of actors at various levels of governance in five case studies categorized under the broad umbrella of green initiatives of India to reveal why some experiments are successful and others are not. Their work provides a comparative analysis of experiments in the case studies to understand how interplay of articulation, participation, and engagement of actors at various levels of the governance structure can only make an experiment successful. The five case studies cover aspects of both mitigation and adaptation.

Chapter 9 by Navarun Varma and Sanjoy Hazarika focuses on the impacts of global climate change on transboundary waters in the context of the Brahmaputra River basin (BRB), an important river system of South Asia. While Brahmaputra can become the engine for economic development in the region, significant natural and anthropogenic challenges hinder progress such as annual floods, erosion, sedimentation, and the incidence of drought. All these challenges cause significant damage to the people, to their livelihoods, and to the overall economy of the region. Despite no historic large-scale water conflict among the Brahmaputra basin countries, growing water demand puts water resources in this basin under pressure (Barua et al. 2017). Recently, climate change and difficult decisions regarding hydropower development have added to these challenges. For example, climate change may increase the amount and intensity of extreme events such as flood and drought. The adaptation activities that upstream countries take in response to such extreme events (such as dams or reservoirs) may impact the abilities of downstream countries to manage the impact of climate change (Blumstein 2016). While there is no dearth of public policy and research focus on ecologically fragile and conflict prone river basins of South Asia, the lack of sustainability in policy and practice is a concern. This chapter uses the conceptual metaphor of institutional fit, as used in sustainability science, to illustrate the mismatches between public policy, problem framing, and social-ecological dynamics in the BRB of the Ganga-Brahmaputra-Meghna interlinked basin. The authors argues that the South Asian water resource management community of practice needs a systemic understanding of the inter-linkages between river basin ecosystems; management sectors for provisioning water, energy, and food; and culture and human aspirations. This chapter concludes with a proposition that current practices of dialogue (for example, Track III dialogue processes for transboundary river basin management), when aided with a transdisciplinary research design to integrate different kinds of logic and stakeholder experiences, can offer opportunities for shared understanding, learning, and co-design of solutions and management pathways.

1.5 Conclusion

Through these chapters and discussed cases, this book seeks to generate new information about the dynamic interactions among actors that shape governance processes and provides a way forward for the academic and the development community. Some of the chapters challenge conventional wisdom. For example, Arabinda Mishra in Chapter 6 notes that we need to pay attention to the emergence and transformation of social capital and not only its erosion as commonly believed. The creation of new forms of social capital may provide avenues for policy makers to embed their efforts in programs for planned adaptation. Similarly, Chapter 5 discusses migration as a positive influence of climate change. The authors argue that factors such as local governance structures, policy responses, and other socio-economic factors rather than climate change mediate migration. The chapter demands more research on the relationship between climate change and migration that takes into account the influence of power and politics at the community level.

At the same time, climate change seems to be a socially constructed process and understanding people's perceptions of climate change on which they base their adaptation responses may be key to the design of new planned adaptation responses. Deliberative governance provides one entry point to making the climate governance processes participatory. It is important to look at climate change along with other stressors such as urbanization which compounds the effects climate change. This book indicates the need for further research to investigate the claims about the popular narratives between climate change and migration and the effects of these processes on social capital. Process documentation research can play a vital role in the efforts to understand community perceptions of climate change and to involve communities in governance mechanisms for adapting to the impacts of climate change.

With these detailed cases and discussions in the context of South Asia, we could identify three major gaps. First, the book was not able to capture the influence of power structures on climate change governance at different levels. Second, as the book discusses the role of various non-state actors in climate change governance, there is a missing link on how and under what conditions non-state actors matter. We argue that non-state actors and their inputs are valuable; however, it is not clear how to use their engagement to influence climate change governance at different levels.

Third, the book has limited focus on mitigation. Although mitigation is an important response strategy to deal with the impacts of climate change, the case studies in this book mostly focus on adaptation because adaptation plays a significant role in South Asia as a response strategy to climate change. The role of adaption is significant because communities realize the benefits of adaptation almost immediately while mitigation is a long-term and expensive option and the realization of its benefits may take decades. Hence, adaptation is a more prominent policy response to climate change in South Asia.

References

Adger, W. N., S. Huq, K. Brown, D. Conway, and M. Hulme. 2003. Adaptation to climate change in the developing world. *Progress in Development Studies* 3(3):179–195.

Ayling, J., and N. Gunningham. 2017. Non-state governance and climate policy: The fossil fuel divestment movement. *Climate Policy* 17(2):131–149.

Barua, A., S. Vij, and M. Zulfiqur Rahman. 2017. Powering or sharing water in the Brahmaputra River basin. *International Journal of Water Resources Development* 1–15.

Blumstein, S., B. Pohl, and D. Tänzler. 2016. *Water and Climate Diplomacy Integrative Approaches for Adaptive Action in Transboundary River Basins*. Berlin: Adelphi, 1–10.

Gemmer, M., A. Wilkes, and L. M. Vaucel. 2011. Governing climate change adaptation in the EU and China: An analysis of formal institutions. *Advances in Climate Change Research* 2(1), doi:10.3724/SP.J.1248.2011.00001.

Huq, S., H. Reid, M. Konate, A. Rahman, Y. Sokona, and F. Crick. 2004. Mainstreaming adaptation to climate change in least developed countries (LDCs). *Climate Policy* 4(1):25–43.

IWMI (International Water Management Institute), 2003. Confronting the realities of wastewater use in agriculture. *Water Policy Briefing* 9. IWMI, Colombo, Sri Lanka.

IWMI, 2006. Recycling realities: Managing health risks to make wastewater an asset. *Water Policy Briefing* 17. IWMI, Colombo, Sri Lanka.

Kuyper, J. W., B.-O. Linnér, and H. Schroeder. 2018. Non-state actors in hybrid global climate governance: Justice, legitimacy, and effectiveness in a post-Paris era. *Wiley Interdisciplinary Reviews: Climate Change* 9(1), doi:10.1002/wcc.497.

Mathur, K. 2008. *From Government to Governance: A Brief Survey of the Indian Experience*. New Delhi, India: National Book Trust.

Narain, V., and A. Prakash (Eds.). 2016. *Water Security in Peri-Urban South Asia: Adapting to Climate Change and Urbanization*. London, UK: Oxford University Press.

O'Brien, K., R. Leichenko, U. Kelkar, H. Venema, G. Aandahl, H. Tompkins, and J. West. 2004. Mapping vulnerability to multiple stressors: Climate change and globalization in India. *Global Environmental Change* 14(4):303–313.

Ojha, H. R., S. Ghimire, A. Pain, A. Nightingale, D. B. Khatri, and H. Dhungana. 2016. Policy without politics: Technocratic control of climate change adaptation policy making in Nepal. *Climate Policy* 16(4):415–433.

Robeyns, I. 2013. *Capability Ethics*. Chichester, NY: Blackwell Publishing.

Vij, S., E. Moors, B. Ahmad, M. Arfanuzzaman, S. Bhadwal, R. Biesbroek, G. Gioli, A. Groot, D. Mallick, B. Regmi, B. A. Saeed, S. Ishaq, B. Thapa, S. E. Werners, and P. Wester. 2017. Climate adaptation approaches and key policy characteristics: Cases from South Asia. *Environmental Science & Policy* 78:58–65.

Wolf, A. T. 2006. Conflict and cooperation over transboundary waters. *Human Development Report Office (HDRO) Occasional Paper 2006–19*. New York: United Nations Development Program (UNDP).

2

"A Disappearing" or a Resilient Ganga? Climate Change Perspectives from the Himalaya

Georgina Drew

CONTENTS

2.1 Climate Change, the Ganga, and Human-Water Relationships 13
2.2 Observing and Understanding Climate Change Phenomena in Uttarkashi District 16
2.3 Flow Variability and Climate Change Perceptions 19
2.4 Flow Variability and Religious Interpretations 22
2.5 Lived Experience and Climate Change Response(s): Some Concluding Remarks 23
References 25

Climate change is associated with rising temperatures, but it will manifest in equally alarming proportions through changes to the quantity, quality, and availability of water. At times the changes to the water balance will result in the scarcity of potable water resources and at other times it will result in the overabundance of water. As the Government of India (GOI) states in its first national document on climate change, the transformation of hydraulic systems should be of grave concern for the climate change adaptation efforts of riparian regions. Efforts to address this transformation should also be of paramount importance for climate change governance. They write: "Changes in key climate variables, namely temperature, precipitation, and humidity, have significant long-term implications for the quality and quantity of water. (The) River systems of the Brahmaptura, the Ganga, and the Indus, which benefit from melting snow in the lean season, are likely to be particularly affected by the decrease in snow cover" (GoI 2008, 16). The GoI additionally notes that this will have an impact on the "total run-off for all river basins" (GoI 2008, 16). To respond proactively, the GoI launched several efforts to manage and mitigate the impacts of changing Himalayan river systems. A strong emphasis of these measures has been on safeguarding the River Ganga, which is the water body that this chapter takes as its concern.

In addition to making the Ganga the "National River of India" and instituting a "National Ganges (Ganga) River Basin Authority" to regulate its "sustainable use and management" (CSE 2008), the Indian central government cancelled a series of dams that were set for implementation near the River Ganga's glacial headwaters. A decision was also made to make the first 100 km of the Ganga's Himalayan flow an Ecologically Sensitive Zone upon which the GoI will not allow large development projects. On paper, these seem like reasonable ways of protecting an important river body that holds religious significance for the hundreds of millions of Hindus who revere its flow as the embodiment of a living goddess. From the vantage of the Himalayan residents who live in the regions where the GoI has instituted these policies, however, elements of these policy measures are questionable. Even as awareness of climate change is growing within the Indian Himalaya, the efforts to govern the Ganga and its surrounding regions do not always fit with the common ways of knowing the river that help to sensitize residents to the threats facing the river that scientific studies highlight. These threats also do not always align with the religion-influenced ways of understanding the River Ganga's properties, including the sentience and omnipotence that residents associate with its waters (see also Drew 2017).

To explain why people do not always embrace some of the policy measures taken to "protect" the River Ganga in the ways that might be hoped, this chapter explores understandings of the Ganga's changing nature(s) from the perspective of residents of the Indian Himalayan zone of Uttarkashi district in an ethnolinguistic region known as Garhwal. The discussion includes familiarity of the residents with the variability and precociousness of the Ganga. The chapter also includes a discussion about how Hindu beliefs influence understandings of the Ganga and its characteristics. To foreshadow what follows, the concerns expressed by the residents of Uttarkashi District demonstrate a distinct split in opinion. On the one hand, the changing dynamics that residents observe lead some to lament that the river Ganga appears to be disappearing (*vilupt hona*). On the other hand, there are residents who believe that no matter what may happen to the regional or even the global ecological balance, the Ganga will remain resilient due to the strength of her divine powers. In both perspectives climate change knowledge impacts the shape of the conversations and discourses. Neither perspective, however, is 100 percent influenced by climate change science. What does appear to be more immediately influential are the risks that major flooding events pose to the well-being of the region and its inhabitants. The concluding remarks, therefore, offer a suggestion that climate change governance could usefully focus on mitigating the flooding hazards that residents might experience soon due to climate change phenomena. This approach does not require contesting the distinct cultural worlds which position the Ganga either as in danger or as self-protecting.

The empirical data outlined in what follows is based on over a decade of research on regional understandings and responses to climate change. This time includes over 18 months spent in field visits to settlements along a tributary running through Garhwal known as the Ganga to devotees (labeled on maps as the Bhagirathi or the Bhagirathi Ganga). These visits occurred in 2004, 2006–2010, 2012, and 2014. The field methods included semi-structured and structured interviews, participation in cultural and religious events, involvement in environmental and social movement campaigns, and the coordination and analysis of surveys about localized perceptions of climate change. This work as well as a large body of scholarship in the physical and social sciences inform the arguments in this chapter. In the latter realm, anthropology and development studies greatly influenced the chapter's argumentation and analyses.

2.1 Climate Change, the Ganga, and Human-Water Relationships

To frame the discussion, it is important to first review the ways that people think that climate change impacts the Ganga. It is also useful to set up how the social science work on human-environment relationships during changing climates helps to understand and approach mitigation and governance efforts.

A prominent concern over the longevity of the River Ganga's Himalayan flow relates to the risks associated with short- and long-term glacial melt. This work is in many ways nascent due to the longstanding difficulties associated with accessing high-mountain glaciers alongside a past dearth of funding for Indian glaciology. Another factor impacting the current knowledge base is that the GoI shrouds in secrecy a good portion of data about India's rivers, and India's glacial balance in particular, due to government concerns about the geo-political sensitivity of this data (Prasai and Surie 2015). However, it is common knowledge that the 30.2 km long Gangotri glacier that feeds the first tributary of the Ganga retreated significantly in the twentieth century. Based on the melting trends of this and many other glaciers, the Intergovernmental Panel on Climate Change (IPCC) gave a "very high" probability that the Himalayan glaciers might disappear by 2035 (IPCC 2007, 493). Although the IPCC withdrew a prediction that the glaciers might soon disappear when select glaciologists disputed the claim (Raina 2009; UNEP 2009), the likelihood of quantitative and qualitative shifts in glacial mass remains. The wealth of scientific data compiled and synthesized by the IPCC and in subsequent scholarly realms indicates that the coming years will witness a rise in global temperatures and a proliferation of dust and black carbon deposits that can impact the integrity of glacial bodies and their ecosystems.

Glacial melt is of course just one of several factors that will impact the water balance of rivers like the Ganga. A changing monsoon and changing precipitation patterns are also points of concern because a good portion of the Ganga's flow comes from the monsoon rains (Basishta et al. 2009). Like the data on glacial melt, however, there are debates among scholars about the severity of this risk. In a review of the literature, Nepal and Shrestha (2015) found that there is no overall change in the precipitation in the Ganges Basin over the past 50 years, adding the disclaimer: "although localized trends have been observed" (p. 207). They note that the projected data indicates that monsoonal precipitation might increase in the future but that the magnitude of the projected change varies depending on the model and study domains employed (Nepal and Shrestha 2015, 207). Other studies contradict this claim. One report indicates that Himalayan rainfall is increasing at a rate of 65 mm per decade, and that the monsoon is getting warmer (Shrestha et al. 2012). More conclusive is the data about the rise of overall temperatures in Himalayan regions that include the Ganga's headwaters (Shrestha et al. 1999). Immerzeel et al. (2012) projected an annual increase in temperatures of about 0.06°C from 2000 to 2100. This increase amounts to a 0.6°C rise in Himalayan temperatures per decade (Qui 2013). The trends indicate that warming in the Himalayan mountains and the Tibetan plateau may be three times greater than elsewhere (Xu et al. 2009). Compelled by the scientific evidence, several scholars have called for the acceleration of climate change mitigation and governance efforts that align with residents' needs and capacities (Dulal 2013). Mohan and Sinha (2015) add to such suggestions by asserting the need for multi-scalar efforts that consider the micro-level and macro-level dynamics of influence for effective climate change responses.[1]

What can be easy perhaps to overlook when looking at the issues from a laboratory or a government office are the cultural systems and religious orientations that influence how people understand the climate change-related threats they face. These cultural systems and religious orientations that influence how people perceive and act in the world are, however, significant to acknowledge and address. As the editors to a volume on *Water, Cultural Diversity, and Global Environmental Change* argue, "Everything is culturally mediated, in all societies" (Johnston et al. 2011, xv). This cultural mediation is apparent in economic activities, politics, and the way people think about and interact with the material environment. The editors of the volume add: "And every social group and every actor in society has a cultural engagement with water. Some of this human/water engagement is manifested in the form of *water culture*: the knowledge, traditional customs, and behavior that support the development and reproduction of a stewardship ethic, or the

[1] Notable for this discussion, Mohan and Sinha (2015) flag Uttarkashi District as one of several north Indian regions that rank as "high" on climate change exposure, sensitivity, and vulnerability indices (Ibid., 316–318). They also state that this region has a very low level of adaptive capacity to the threats and vulnerabilities that they face (Ibid., 320).

political organization of societies to manage and maintain water resources" (Johnston et al. 2011, xv). If the consensus is that this cultural mediation is to a large degree true, then it follows that efforts to understand social responses to threatened bodies of water such as the Ganga must engage the cultural worlds in which these rivers hold significance.

A growing body of work in the anthropology of climate change aims to underscore the role of culture by using grounded studies of people's daily lives to understand how they perceive and respond to climate change phenomena and their associated risks. To do this, anthropologists often use ethnography, which is a process that entails short- and long-term periods spent living with people in a given setting while interviewing them about their beliefs, actions, and desires or fears. Ethnography also involves looking at, and working to understand, the particular yet messy and complex lives that people lead (Abu-Lughod 2000). In 2009, the landmark book *Anthropology and Climate Change: From Encounters to Actions* (Crate and Nuttall 2009) propelled anthropological efforts to ethnographically document socio-cultural responses to climate change. The book and its editors contended that climate change threatens cultural systems, including patterns of existence in and perception of the world. The editors argued that global ecological change is increasingly challenging culturally diverse and enriching human-environment relations. These relations lose place because climatic shifts entail the relocation of human, animal, and plant populations as they adjust to change (Crate and Nuttall 2009, 12).

As the scholarship contends, such relocations entail a loss of intimate human-environment relationships that not only ground and substantiate indigenous worldviews, but also work to maintain and steward local landscapes (Crate 2008; Crate and Nuttal 2009). In some cases, migrations and ecological change will also result in the loss of mythological symbols, meteorological orientation, and even the very totem and mainstay plants and animals that ground a culture (Crate and Nuttall 2009, 12).

It is these risks along with the threats to the physical environment that such researchers argue need consideration when working to understand how people respond and why (Crate 2008). Researchers can do this while recognizing that the cultures, like environments, are subject to transformation over time.[2] The chapter illustrates ways of knowing in the following sections.

[2] As Johnston et al. (2011) remark: "All humans have learned behaviour expressed in the patterns we call culture. These learned patterns of behaviour serve many functions, not the least of which is the ability to interact, engage, and adapt to changes in the environment and in society. ... While all cultures evolve and change, the means to support and sustain a cultural way of life may be destroyed, and this destruction can have profound impacts on the ability to utilize and reproduce culturally-distinct knowledge, values, and traditions" (xiv).

2.2 Observing and Understanding Climate Change Phenomena in Uttarkashi District

The ways of knowing that are important for this discussion are understandings of the Ganga's daily and seasonal fluctuations as well as, to a lesser degree, ways of understanding climate change. Knowledge about the Ganga's flows and variations take prominence in this discussion because the river's dynamics of change are obvious to residents, and can therefore be queried. For many mountain residents, climate change remains incomprehensible and the difficulty of reaching a consensus on what the term means complicates questions about the manifestation of climate change. Understandings of changes to the water balance, however, are easier to ascertain. And since changes to the water balance are a feature of climate change phenomena, these changes can extrapolate back to how people respond to transformations in their localized environments and what they relate to climate change. Another reason for focusing on water is that it is also a good measure for thinking about human-environment relationships. Water, after all, is "simultaneously an element, a flow, a means of transport, a life-sustaining substance, and a life-threatening force" that inspires novel ways of thinking about social relations and human relationships with the environment (Krause and Strang 2016, 633–634).

While residents in the Indian Himalaya observe manifestations of change on and along the Ganga, these changes are not always obvious to passing visitors. The first reason for this is that seeing a river does not necessarily equate to the understanding that is enabled by long-term observation. The second reason for this is that perceptions and interactions with the Ganga that draw from the Hindu faith influence experiential and phenomenological knowledge, which is part of what Ingold (2010) calls a dwelling perspective. These dynamics have an impact on perceptions of how people should or could manage the river.

To restate the first reason, it takes time to know a river. A river has its own rhythms and its own cycles. It has sounds that are specific to its topography and to the rocky or sandy materials over which it passes. A river even has scents—from sweet to sour and acrid—that distinguish it for those who have taken the time to know it well enough. And some rivers, such as the Ganga, bestow spiritual and religious blessings upon their observers. Rivers of this latter type may also feel, think, and act; they may be knowers themselves. These multiple modes of knowing are—like the characteristics of water—ever changing. After all, and as a leading scholar on water and culture writes, "Like water, knowledge is always in fluid motion: trickling and seeping, permeating, flooding and swamping…" (Strang 2015, 52). The rhythmic, sonic, sensorial, spiritual, and agential is all part and parcel of knowing a specific tributary of the River Ganga that passes through a stretch of the Indian Himalaya known as Garhwal.

To restate the second reason, people often perceive daily observations about change in rivers such as the Ganga through the prism of religious beliefs and practices. Garhwal residents and devotees associate personhood with this liquid flow, and they consider the river to be a knower as well as an entity for them to know. Adding to the respect and relationality that Garhwalis feel for the Ganga are the numerous stories about the spiritual strength of this river's flow as it moves through the Himalaya. The Hindu epics, such as the Mahabharata and the Ramayana, say that the Ganga, the Goddess, offered many boons to mortals, sadhus, and saints traveling in the mountains. These texts portray the Goddess as a beautiful, sensuous maiden who can step out of the waters to take on human form whenever she desires. Her purpose on earth, those texts go on to explain, is to purify the devotees who bathe in her waters from their sins and to serve as a bridge to the divine for living and departed souls. The faith associated with the Ganga's flow demonstrates how water flows "through religious beliefs as well as political, economic, and social practices" (Strang 2015, 7).

The difference between seeing and knowing was a distinction that I realized over a period of repeated visits to the Ganga's Himalayan banks from 2004 to 2014. The first time I saw the Ganga's flow through Uttarkashi was in November 2004 when I was passing through the district capital for a trek up to the river's glacial source about 125 km upstream. I recall sitting down alongside the river as it turned a curve in the topography at a spot just upstream from the town's main market. As I invoke this sight in my mind's eye, the river is at a "Goldilocks level"—it is neither so full as to be dangerous, nor so shallow as to be unimpressive; the river, as I think back on that moment, was just right. When I returned for preliminary fieldwork for an ethnographic project in July 2007, however, the river seemed transformed. Gone were the clear, crisp waters of the winter flow. In their place was the furious rushing and gushing of grey, sand-and-mud filled monsoonal waters. At the time, the river was so wide and so fierce that it was only at locations known as *ghat*, which are a series of concrete steps leading to the riverbank, where devotees could safely perform their prayers and ritual acts of worship. At one specific *ghat*, the river was so fierce that people had installed a series of metal poles at the bottom of the steps with a series of two-tiered chains strung from pole to pole. Devotees cling to these chains when the Ganga threatens to sweep them away as the water races around their ankles.

As time wore on, a new manifestation of the river appeared on every visit. It was with a shock that, when arriving in the dry months of January to March, I discovered that the river had a third form. Along with the crystal clarity of the just-right post-Monsoon flow and the rushing-gushing of the mid-Monsoon flow, the Ganga had a barely-there flow. In this last form of the river's cyclical manifestations, the main channel was little more than a rivulet. On the edges of the banks, standing water grew stagnant and turned green with the accumulation of algae. These stagnant algae-filled waters are what greeted the eye upon emerging from the houses and temples built along the river's banks. To find the still-flowing rivulet required crossing a veritable sea of pebbles, rocks, and boulders to the middle of the riverbed. At such times, it became a hydrological game of hide-and-seek to access the Ganga's flow. This, of course, is what outsiders perceive when seeing the river in fleeting, short-term encounters. It is an altogether different thing, according to residents, to live alongside the Ganga's flow and to have knowledge of its movements and moods. As Uttarkashi residents stated, they are already quite familiar with the wide fluctuations of which the Ganga is capable. These fluctuations are not necessarily recent phenomena as tales from older generations attest to times in decades past when the Ganga flooded unexpectedly or reduced to a small trickle (Figures 2.1 and 2.2).

FIGURE 2.1
The Ganga's flow through Uttarkashi, July 2007. (Courtesy of Georgina Drew.)

FIGURE 2.2
The Ganga's flow through Uttarkashi, April 2009. (Courtesy of Georgina Drew.)

2.3 Flow Variability and Climate Change Perceptions

The Uttarkashi-based perspective initially became apparent in casual conversations and semi-structured interviews with interlocutors who I came to know during a research project that initially involved over a year of living in Uttarkashi from 2008 to 2009 with additional periods of observation documented in subsequent years. It was during this timeframe that the Ganga's enchantingly chameleon characteristics became more apparent. It was also through the process of ethnographic fieldwork that I came to reflect more frequently on the insights of scholars such as Hugh Raffles (2002) and Doreen Massey (1994) who speak of the ever-changing and evolving nature of place. Raffles, in his study of an Amazonian river, is particularly attentive to how humans and non-humans (e.g., rivers) are involved in acts of creativity as they transform the landscapes that they inhabit. And while the places they co-create may look static when seen in passing, they are ever-evolving. As Raffles explains: "It is the impression of stasis that beguiles. They may look secure, but landscapes are always in motion, always in process" (Raffles 2002, 34). As for the Igarapé Guariba, the river that he studied and lived along, he speaks of its force, and hence its creative potential, as so excessive that it commands recognition (Raffles 2002, 34). "The river," he writes, "will not allow you to ignore it. The land shifts of its own accord. The banks crumble, the fields flood, the orchards float off to the horizon" (Raffles 2002, 35). This watery agency, Raffles argues, means that researchers have to situate the study of rivers within "that other form of fluvial practice" that is "the genealogical… the space where facts get made out of history, theory, and strategy" (Raffles 2002, 35). And it is the intertwining of non-human fluvial practices with human fluvial practices (and engagements) that Raffles ends up terming "fluvial intimacies" (Raffles 2002, 180–184).

The many ways of interacting with the full and rushing Ganga preclude just one glimpse of human-river interactions. It is common for people to arrive, touch the waters reverentially, apply the water to the forehead and the top of the head, and then fold the palms of their hands together in greeting and prayer. Some people walk away after doing this, though they might also cup some water into their hands and take in a deep sip—never mind the grit of the monsoonal flow that has brought with it the silt and soil of upstream regions. Other people take longer and perform more elaborate rituals. They might light small candles or incense (which they invariably place on the tops of the poles installed to keep them safe or on a patch of concrete near their feet). In Uttarkashi, women have particularly elaborate rituals. Some of these rituals involved taking out large trays filled with spices, edibles, and beauty enhancements. These women would take pinches of these elements and throw them into the roaring gush of brown-grey liquid in front of them. When I asked one woman why she offered rice and spices to the Ganga in this manner, she retorted back sharply, "Why do you eat?!" The rebuttal was so sharp I did not bother to ask why she had also offered the red *kukum* paste to the Ganga, as well; the woman was sure to respond to such a question with a retort demanding to know why I also wear makeup. Her intention was to point out the agency and personhood of the entity before us.

When the Ganga is rushing past your ankles and threatening to pull you into the racing current, it is easy to feel her power. It can feel playful, but it can also be dangerous. The energy of this non-human being can be so excessive in the times of monsoonal flow that it not only forces recognition (per the earlier quote by Raffles 2002, 35), but it also exacts fear and awe. Experience of the river's wild movement also informs regional understandings that the Ganga can move and change course. This movement, reportedly, is something that the river has done many times in the past. As a schoolteacher and Ganga devotee told me, the Ganga used to flow on a completely opposite part of Uttarkashi. At the time of the interview, the river flowed south of the township; moving in this direction, the Ganga rounded a bend that moved in a half-circle to the northern side after passing by the main market. Previously, perhaps over a hundred years ago, the Ganga merely cut through the opposite side, racing in a straight line from one end of the town to the other on her way down the mountains. As the schoolteacher declared: "The river changes of her own accord… From one place to another she moves… she changes."

Possibly, considering the Ganga's capriciousness, companies and government organizations would steer clear of building dams on such turbulent waters; however, this was not the case until a series of floods through the region took thousands of lives over the summers of 2012 and 2013. An expert review of dam building that followed recommended the cancelation of 23 dams in the region, although the debate about their feasibility continues. Possibly river devotees would object to the imposition of dams on a powerful and religiously important entity, and that they might lament the ways that such projects harm the Ganga's potency; however, this was occasionally, but not always, the case. In fact, it was the belief of the devotees in the Ganga's powers that led them to believe that dams could not harm the river and its Goddess. As the schoolteacher stated in the interview, "There is a strength (*takat*) in her; and that strength will never be diminished." The assertion of the Ganga's never-ending strength was of interest because at the time the discussion was on glacial melt and the projections of water shortage in the Himalaya that could manifest as a result. The discussion also touched upon the impact of the dams upstream—several of which were still under construction at the time of the interview. Speaking of these issues while maintaining a stance that the Ganga will retain her force and her strength no matter what happens, my interlocutor made the following statement:

> In my opinion, her condition has not (yet) changed. Though it has come to pass that because of dams, and from building everything else, there has been a little alteration… Maybe with modernity's 'development' and 'technology' we have had to pay a little price. And we have to pay it. If you build a road… you have to cut the mountains. It is only in that way that roads are built; you can't make a road in the air. So, if you are going to make a dam, if you have to make electricity, then you will have to send her into another route. But my feeling is that in doing so, not so much 'change' happens as people portray it. But… yes, there is some alteration.

In this statement, the schoolteacher makes a distinction between change and alteration. The reason for this emphasis is that the schoolteacher sees the divine force of the Ganga as more powerful than the impact that humans can have on the river and its Goddess. So, humans may alter the river, but the river retains an essential core that is powerful no matter what. This, interestingly, overlaps with the insights of Alley (1998, 2002) on reactions to pollution in the Ganga. According to her research in Varanasi, devotees can observe polluting elements in the river, but they do not necessarily consider that the debris diminishes the Ganga's divine power and purifying abilities.

2.4 Flow Variability and Religious Interpretations

The schoolteacher's conviction of the Ganga's inalterable strength is also due to Hindu beliefs that the Ganga comes from a divine heavenly source. When she descended thousands of years ago from the heavens to the Earth at the bequest of a mortal, King Bhagirath, the Ganga brought a benediction to the world. And, along with her presence in heaven and on the Earth, the Ganga also flows in a third realm, the netherworld known in Hindi as *patal* (Mankodi 1973, 139). The name *tripathaga* for the Ganga means "she who moves through three courses" of three worlds (Mankodi 1973, 139). Whereas the Ganga's physical form may rise or shrink in the world of mortals, she remains strong in the heavens and in the netherworld. Teachings about the Ganga's origins and characteristics in Hindu epics strengthen these beliefs.

Another interlocutor helped to explain the significance of these understandings for how devotees perceive change in the Ganga. This man, a part-time Brahmin priest who is also an engineer, stated that it was because of the Ganga's ability to flow on multiple plains that she will never leave the region of Garhwal. This comment was given in a focus group session in the summer of 2009 and it was in response to a female participant's concerns that the Ganga might one day disappear. The Brahmin quickly interjected, "It is not like that… (It is not) that she will never be there. She is on the ground now, flowing, but she is also underground, in *patal*. She will never be gone. But she may go into *patal*." He added: "The Saraswati River is also gone from the earth. She has gone underground, into *patal*. Once she goes, she can't come back."

When commenting on the eventual inevitability that the Ganga will one day go underground, the Brahmin stressed that the process could take hundreds or thousands of years. He also declared that even as the Ganga moves underground, her devotees' faith would remain. As he noted, people still pray to Saraswati even though they cannot access the Goddess' physical waters. According to such interlocutors, people still receive blessings and boons from "Mother Saraswati" if their faith is true. This, he said, may one day be the fate of the Ganga, or "Mother Ganga." The implication of this prediction is also ambivalent in nature: the Ganga will move from one location to another—from the land or into the ground.

Another reason for Hindu beliefs in the Ganga's continued existence is that the Goddess Ganga embodies the divine feminine energy that is ultimately a part of every Hindu Goddess. While the Ganga is a distinct entity, she is also a symbol of a sacred feminine force that characterizes all Goddesses. Tracy Pintchman (1994) calls this force the "Great Goddess" and she attributes to it an impersonal cosmic power that transcends the particularities of specific deities. In almost all her manifestations, the Great Goddess is "[a] benevolent provider of grace to her devotees" (Prasad 2003, 254). It is because of the numerous powers and qualities associated with the Great Goddess that devotees represent the Ganga in an anthropomorphized form, such as that of a woman sitting upon a crocodile, while also revering and worshiping her in a riverine state. These forms help devotees remember the Goddess' power, or *shakti*, as well as her role in nature, or *prakriti*. Deeply intertwined, *shakti* and *prakriti* are integral to the understanding of the Goddesses' grace. These are characteristics that devotees behold when looking upon the Ganga's waterscape that nonbelievers may not so readily perceive.

A point to underscore is that a Goddess such as the Ganga not only has a *shakti* that is independent of her physical riverine form, but also has a part of nature that mortals can access to seek her blessings. And just because her physical manifestation, her *prakriti*, might appear diminished, it does not mean that her power, her *shakti*, is diminished. These distinctions influence how devotees make sense of changes (or alterations) in the Ganga's flow. It also helps temper some their fears (depending on the person) that development and ecological change will harm the Goddess residing in the Ganga.

2.5 Lived Experience and Climate Change Response(s): Some Concluding Remarks

The forms of knowing the River Ganga and its Goddess described in this chapter involve a suite of embodied and sensorial experiences alongside personalized interpretations that extrapolate to climate change phenomena. As with other rivers, the forms of knowing include understanding the river's movements developed over time (Krause 2013). This understanding is based on observations of the rising and waning of the Ganga's waters, as well as on the capriciousness of the river's nature when the Ganga moves suddenly to reclaim land or to destroy the settlements that come across her wrath as she surges and floods. While the Ganga is beginning to move in ways difficult to predict, there are explanations for this that do not necessarily relate to climate change. The changes, as some residents contend, could be due to the implementation of dams on the river's waterways or even due to the wild fluctuations of which the river has been capable for several decades. And as devotees note, the changes could also be due to the Ganga's own volition; it is possible that she is retreating underground, as did the Saraswati River dozens of centuries ago (Danino 2010), and as the Hindu texts predict the Ganga to do in the future (Nautiyal 1994). This understanding demonstrates that devotees draw upon "mythology as a long-standing resource" to make sense of the changes to the river's form (Khan 2016, 182). Diverse interpretations of the Ganga's permeations and transmutations demonstrate how people perceive and interact with changes in their aquatic environments.

All the discussed reasons point to why residents of the Himalaya temper some of the climate change science when they observe the changes in their localized environments. For this reason, Garhwali residents might react strongly against efforts to create a conservation zone along the river's first 100 km. They may view such policies as a threat to the development prospects of the region (Tripathi 2013; Upadhyay 2017). Residents of the Garhwal Himalaya can similarly greet other policy measures, such as the Uttarakhand government's efforts to declare the Ganga a "living person" with nonplussed reactions; these moves can appear to only underscore what they already know about the river's personhood and agency (Daley 2017; Lokgariwar 2017). Personhood declarations can even become suspect because they tie into the conservation ethic that some perceive might prioritize the wellbeing of the River Ganga over the welfare of the people living alongside the river.

Rather than crafting policy to protect the Ganga, therefore, a policy approach that has more scope for finding common ground is one that focuses on protecting the lives and livelihoods of Garhwali residents (Drew 2014, 2017). A focal point of agreement, for example, could be the need for efforts to protect the region and its inhabitants from the severe flooding events such as the ones that took place in the summers of 2012 and 2013. These massive floods displaced thousands and killed at least 5,700 people (Qiu 2013). Even the residents of Garhwal who are either ignorant or sceptical of the science linking such events to climate change would agree that they need proactive efforts to expand the safeguards to protect residents and their properties from future disasters. Part of these efforts might involve educating people about the mechanisms of past floods while alerting them to the likelihood of future flooding events. This effort could help to gain the support of Garhwali residents for revised development plans that avoid constructing dams and buildings near to the Himalayan glaciers or the river's capricious flow. Such measures would have the result of feeding into the wider climate change mitigation and governance goals that the GoI has outlined while making residents feel that their wellbeing is the priority. This work could advance with the aim of further calibrating regional perceptions of risk with the logics of the diverse cultural worlds that shape the lives of Garhwal's Hindu-identified populations. The GoI needs efforts along these lines because, as other scholars have suggested, the study of how social and hydrological relationships are interconnected helps to give a much deeper understanding of the role of water in human social lives so that "significantly better management and policy can be designed" (Krause and Strang 2016, 633).

To underline some key points from this discussion, culturally influenced and local contingent observations of change do not always align with the assertions of an ever-evolving climate change science. At the same time, top-down policy prescriptions do not always align with the perspectives and concerns of the populations affected by such policies. These observations suggest that there is a need to craft people-centered and culturally appropriate measures that simultaneously feed into governance aims and targets. Whether one believes that the Ganga is disappearing or resilient, there are points of agreement in the immediate present that researchers can use to advance proactive climate change governance measures.

References

Abu-Lughod, L. 2000. Locating ethnography. *Ethnography* 1(2):261–267.

Alley, K. 1998. Idioms of degeneracy: Assessing Ganga's purity and pollution. In *Purifying the Earthly Body of God: Religions and Ecology in Hindu India,* L. E. Nelson (Ed.), pp. 297–329. Albany, NY: State University of New York.

Alley, K. 2002. *On the Banks of the Ganga: When Wastewater Meets a Sacred River.* Ann Arbor, MI: University of Michigan Press.

Basishta, A., D. S. Arya, and N. K. Goel. 2009. Analysis of historical changes in rainfall in the Indian Himalayas. *International Journal of Climatology* 29:555–572.

Centre for Science and Environment (CSE). 2008. *Will Ganga Get Its Life Back?* http://www.cseindia.org/misc/ganga/ganga_action.htm (accessed May 15, 2009).

Crate, S. A. 2008. Gone the bull of winter: Grappling with the cultural implications of and anthropology's role(s) in global climate change. *Current Anthropology* 49(4):569–595.

Crate, S., and M. Nuttall. 2009. Introduction: Anthropology and climate change. In *Anthropology and Climate Change: From Encounters to Actions,* S. Crate and M. Nuttall (Eds.), pp. 9–36. Walnut Creek, CA: Left Coast Press.

Daley, J. 2017. India's Ganges and Yamuna rivers are given the rights of people. *Smithsonian Mag,* March 23. https://www.smithsonianmag.com/smart-news/ganges-and-yamuna-rivers-given-rights-people-india-180962639/ (accessed November 2, 2017).

Danino, M. 2010. *Lost River: On the Trail of the Sarasvati.* New Delhi, India: Penguin Books.

Drew, G. 2014. Developing the Himalaya: Development as if livelihoods mattered. *Himalaya: A Journal of Nepal and Himalayan Studies* 34(2):31–38.

Drew, G. 2017. *River Dialogues: Hindu Faith and the Political Ecology of Dams on the Sacred Ganga.* Tucson, AZ: University of Arizona Press.

Dulal, H. B. 2013. Governing climate change adaptation in the Ganges basin: Assessing needs and capacities. *International Journal of Sustainable Development & World Ecology* 21(1):1–14.

Government of India. 2008. *National Action Plan on Climate Change.* New Delhi, India: Prime Minister's Council on Climate Change. http://pmindia.nic.in/climate_change.htm (accessed October 20, 2017).

Immerzeel, W. W., L. P. H. Beek, M. Konz, A. B. Shrestha, and M. F. P. Bierkens. 2012. Hydrological response to climate change in a glacierized catchment in the Himalayas. *Climatic Change* 110(3–4):721–736.

Ingold, T. 2000. *The Perception of the Environment: Essays in Livelihood, Dwelling, and Skill.* London, UK: Routledge.

International Panel on Climate Change (IPCC). 2007. *Climate Change 2007: Climate Change Impacts, Adaptation, and Vulnerability, Summary for Policy Makers.* Contribution of the Working Group II to the Fourth Assessment Report of the Intergovernmental Panel on Climate Change, February 2007. Brussels, Belgium: IPCC.

Johnston, B. R., M. Barber, V. Strang, I. Klaver, L. Hiwasaki, and A. R. Castillo. 2011. Preface. In *Water, Cultural Diversity, and Environmental Change—Emerging Issues, Sustainable Futures?* B. R. Johnston, V. Strang, I. Klaver, L. Hiwasaki, and A. R. Castillo (Eds.), pp. xi–xxi. New York: Springer.

Khan, N. 2016. Living Paradox in Riverine Bangladesh: Whiteheadian perspectives on Ganga Devi and Khwaja Khijir. *Anthropologica* 58(2):179–192.

Krause, F. 2013. Rapids on the 'Stream of Life': The significance of water movement on the Kemi River. *Worldviews* 17(2):174–185.

Krause, F., and V. Strang. 2016. Thinking relationships through water. *Society & Natural Resources* 29(6):633–638.

Lokgariwar, C. 2017. *The Sad State of These Persons Called Ganga & Yamuna—Can (the) State Protect Them?* https://sandrp.wordpress.com/2017/04/11/the-sad-state-of-these-persons-called-ganga-yamuna-can-state-protect-them/ (accessed November 2, 2017).

Mankodi, K. 1973. Ganga tripathaga. *Artibus Asiae* 35(1/2):139–144.

Massey, D. 1994. *Space, Place, and Gender.* Minneapolis, MN: University of Minnesota Press.

Mohan, D., and S. Sinha. 2015. Assessing vulnerability to climate change in the Ganges basin using a combined macro- and micro-level approach. *Climate and Development* 8(4):312–323.

Nautiyal, S., trans. 1994. *Skand Puran.* Allahabad, India: Hindi Sahitya Sammelan.

Nepal, S., and A. B. Shrestha. 2015. Impact of climate change on the hydrological regime of the Indus, Ganges and Brahmaputra basins: A review of the literature. *International Journal of Water Resources Development* 31(2):201–218.

Pintchman, T. 1994. *The Rise of the Goddess in the Hindu Tradition.* New York: State University of New York Press.

Prasad, L. 2003. Hindu Goddesses. In *South Asian Folklore: An Encyclopedia: Afghanistan, Bangladesh, India, Nepal, Pakistan, Sri Lanka,* M. A. Mills, P. J. Claus, and S. Diamond (Eds.), pp. 254–259. New York: Routledge.

Prasai, S., and M. D. Surie. 2015. Water and climate data in the Ganges basin: Assessing access to information regimes and implications for cooperation on transboundary rivers. *Water Alternatives* 8(2):20–35.

Qiu, J. 2013. Floods spur mountain study: Himalayan nations take action in response to changing climate and its deadly effects. *Nature: International Weekly Journal of Science* 501:14–15.

Raffles, H. 2002. *In Amazonia: A Natural History.* Princeton, NJ: Princeton University Press.

Raina, V. K. 2009. *Himalayan Glaciers: A State-of-Art Review of Glacial Studies, Glacial Retreat and Climate Change.* Government of India, Ministry of Environment and Forests Discussion Paper. http://moef.nic.in/downloads/public-information/MoEF%20Discussion%20Paper%20_him.pdf (accessed February 2, 2010).

Shrestha, A. B., C. P. Wake, P. A. Mayewski, and J. E. Dibb. 1999. Maximum temperature trends in the Himalaya and its vicinity: An analysis based on temperature records from Nepal for the period 1971–94. *Journal of Climate* 12(9):2775–2786.

Shrestha, U. B., S. Gautam, and K. S. Bawa. 2012. Widespread climate change in the Himalayas and associated changes in local ecosystems. *PLoS One* 7(5):e36741. doi:10.1371/journal.pone.0036741.

Strang, V. 2015. *Water: Nature and Culture*. London, UK: Reaktion Books.

Tripathi, P. S. 2013. Politics of ecology. *Frontline*, June 14. http://www.frontline.in/environment/conservation/politics-of-ecology/article4756368.ece (accessed November 2, 2017).

United Nations Environment Programme (UNEP). 2009. *Recent Trends in Melting Glaciers, Tropospheric Temperatures over the Himalayas and Summer Monsoon Rainfall over India*. na.unep.net/siouxfalls/publications/Himalayas.pdf (accessed October 21, 2017).

Upadhyay, K. 2017. Not so quite flows the Bhagirathi in Uttarkashi. *The Hindu*, February 5. http://www.thehindu.com/elections/uttarakhand-2017/Not-so-quiet-flows-the-Bhagirathi-in-Uttarkashi/article17195577.ece (accessed November 2, 2017).

Xu, J., E. R. Grumbine, A. Shrestha, M. Eriksson, X. Yang, Y. Want, and A. Wilkes. 2009. The melting Himalayas: Cascading effects of climate change on water, biodiversity, and livelihoods. *Conservation Biology* 23(3):520–530.

3

Deliberative Governance on Vulnerability to Climate Change: Voices from Madhesi Farmers

Floriane Clement

CONTENTS

3.1 Introduction 29
3.2 Exploring Deliberation in the Global South 31
3.3 Methodology 34
  3.3.1 Research Area 34
  3.3.2 Action Research 35
  3.3.3 Organization of the Public Dialogues 36
  3.3.4 Data Analysis 37
  3.3.5 Limitations 38
3.4 Contrasting Framings of Vulnerability 38
  3.4.1 Climate Change Policies in Nepal 38
  3.4.2 Framings of Vulnerabilities in Farmers' Films 40
3.5 A Step Towards Deliberative Governance 42
  3.5.1 VDC Level: Farmers Only 42
  3.5.2 District Level: Farmers and Government Line Agencies 43
  3.5.3 National Dialogue: Farmers, Civil Society, and Government Officials 45
3.6 Discussion and Conclusion 46
Acknowledgments 47
References 47

3.1 Introduction

The design and negotiation of much of the climate change adaptation policies and programs are outside the reach of ordinary citizens. This situation is especially true in the rural South where international agencies heavily influence and fund climate change adaptation policies and interventions

(Arndt and Tarp 2017). The United Nations Framework Convention on Climate Change (UNFCCC) provided a framework for least-developed countries (LDCs) to develop a National Adaptation Program of Action (NAPA) through a participatory process to identify priority activities and projects for adaptation. Although policy intentions are to include local and contextual framings of climate risks, some scholars note the inherent tension between the global environmental risks perspective embedded in the UNFCCC and the ability of the NAPAs to open sufficient space to integrate alternative framings (Ayers 2011). Climate policies and interventions in the South largely have relied on supposedly universal "good" adaptation practices a priori identified by experts and scientists (such as the adoption of drought-resistant seed varieties) and the creation of climate-smart infrastructures and systems (such as water-storage reservoirs or crop insurance).

There has been an increasing recognition that climate change adaptation programs also need to be locally relevant. Integrating local values, perceptions, and knowledge in their design and implementation is important for several reasons. First, the effects of increased climatic risks and variability can significantly vary locally, and the specific knowledge local people have of the social-environmental system in which they live can help in identifying and assessing these variations and their causes (Lebel 2013). Second, these effects interact with several global to local factors from economic globalization (O'Brien and Leichenko 2000), national political economic structures and pre-existing social inequities (Merino and Ribot 2012; Yates 2012), to local culture, social norms, and biophysical environment (Jones and Boyd 2011). Not considering local contextual factors might result in unanticipated policy effects which could lead to increased vulnerabilities of the most marginalized. Last, local people have an intimate knowledge of their own vulnerabilities through their physical and mental experiences and their own understanding of the causes of their vulnerabilities (Eriksen et al. 2011). Documenting such intimate and often non-discursive knowledge will help in understanding what various people value in their lives, based on their identities, social groups, and geographies.

The importance of integrating local view and knowledge in the development of climate change adaptation policies and programs is therefore not only a question for global climate change adaptation efforts of finding "the right fit" to local social-ecological systems (Byg and Salick 2009). It is also a question of how legitimate and relevant global institutions and initiatives are in the eyes and lived realities of the people they target, and how their interaction with political economic structures and existing vulnerabilities might aggravate or reproduce inequities and mechanisms of exclusion (Clement 2013; Eriksen et al. 2015; Merino and Ribot 2012; Smucker et al. 2015).

The recognition that climate change adaptation policies should be locally grounded materialized in the call for the participation of local communities in identifying and prioritizing suitable adaptation activities in the preparation of NAPAs. However, the question is to what extent such participatory processes can lead to improved climate change governance. In the analysis of the NAPA in Bangladesh, Ayers (2011) notably remarks that such programs might fail to address critical questions on who participates, how is participation orchestrated, and whose knowledge and expertise are privileged. Such issues resonate with the context of several other countries in the global South that are characterized by elite-dominated bureaucracies and low downward accountability of functionaries and politicians. Whereas several scholars have called for deliberative forms of governance and policy-making in natural resource management (Dore 2014; Forsyth 2011; Zachrisson 2010), there has been little attention on how to initiate real forms of deliberation in settings dominated by severe power imbalances among stakeholders (bureaucrats, farmers, non-government organization (NGO) representatives, etc.), high divergence of views and interests, and lack of democratic processes (Neef 2009). This is the research challenge addressed in this chapter which proposes a critical reflection on the relevance of deliberative forms of governance in such contexts. Our research question was the following: to what extent can deliberation open the discursive space to integrate local framings of global environmental risks in national debates?

The process we initiated in Nepal was an action research project implemented over a period of 2 years which culminated in a series of deliberative dialogues on vulnerabilities that gathered together farmers, policy-makers, experts, and civil society representatives. We examined the extent to which deliberation practices supported the development of a shared problem framing, guided by the belief that "governance is seen to be as much about shared problem construction as it is about collective solutions" (Leach et al. 2007, 28). We also aim at bringing fresh empirical insights to the body of knowledge that draws on interpretive approaches to climate change governance (Methmann et al. 2015; Stevenson and Dryzek 2015).

The chapter starts with a reflection on the use of deliberative tools in the global South in contexts often characterized by an unequal public sphere. Following an overview of the methodology for this action research project, the next section sets the scene with a short presentation of climate change policies in Nepal and the story lines prominent in farmers' discourses. The results section presents the analysis of discourses in the public debates followed by a discussion and conclusion section on the general lessons drawn from the research.

3.2 Exploring Deliberation in the Global South

Globally, the call is for greater citizen involvement in policy processes, notably in the perspective of enhancing environmental and social justice. This call is based on the premise that policy-makers are not necessarily better able than ordinary citizens to define problems and solutions. This view also reflects the post-empiricist approach whereby experts primarily do not produce knowledge but citizens through deliberative practices voice competing views (Fischer 2003).

The theory of communicative action by Habermas (1984) has influenced heavily the underpinnings of theories of policy learning and the need for deliberative decision-making whereby deliberative practices can help a group reflect on its own interests and reframe them. By deliberation, we mean a process of face-to-face-engagement whereby people "carefully examine a problem and arrive at a well-reasoned solution after a period of inclusive, respectful consideration of diverse points of view" (Gastil and Black 2008, 8). True forms of deliberation are based on reasoning and persuasion and exclude other power mechanisms such as coercion, deception, or manipulation (Dryzek 2000). However, more subtle and dispersed forms of power might be at play during deliberative processes. Discourses and institutions exercise power by giving legitimacy to specific forms of knowledge (e.g., written forms, expert knowledge, knowledge from certain disciplines), thereby rendering certain voices and actors invisible. These forms of power, particularly evidenced in Foucault's work (Foucault 1975), are not vested in specific individuals and not necessarily used intentionally through the exercise of agency. Deliberation, therefore, might not be sufficient to challenge the status quo when powerful actors perceive experts' knowledge as the only valid and legitimate form of knowledge.

A core assumption under the call for deliberative governance is that "various groups of people conceive of the world in different terms" (Hajer and Wagenaar 2003, 11). For post-empiricists, language and discourses are central in framing social and environmental problems—in turn delineating who can discuss solutions and how—and in creating actors' coalitions around shared story lines (Hajer 1995). Story lines, a sort of narratives, are important discursive and political devices. They allow actors to position themselves around specific narratives which establish causalities between certain socio-environmental phenomenon (e.g., land degradation, vulnerability to climate change, etc.) and a specific set of factors and processes. These discursive coalitions—consisting of story lines and of the actors supporting these story lines—play a key role in policy change by assigning meaning and values to environmental problems, often operating discursive closure and implicitly giving legitimacy to certain solutions rather than others (Hajer 1995; Roe 1991).

Plural framings are contingent upon social values, economic interests, and organizational structures, which make it difficult to separate facts from values (Forsyth 2003). How participants interactively and mutually negotiate different framings has an important bearing. In a deliberative governance process, participants test the validity of any one interpretation against earlier interpretations and accept or reject them through dialogue among participants. The next step involves a discursive synthesis of competing views and recognizing dissent as a legitimate discursive contribution. The objective, therefore, is not for the participants to achieve consensus but to acknowledge that there are differing views and that others' views are valid and that participants should consider all views for joint problem solving.

Gastil and Black (2008) propose a framework with a set of principles for effective deliberation. They distinguish the analytic process, which outlines the functional components and steps of deliberation such as "create information base," "prioritize key values," "identify solutions" and so on, and the social process, through which deliberation occurs. The latter is based on mutual comprehension, speaking opportunities, consideration, and respect (Gastil and Black 2008). For each of the steps of these processes, they propose an appropriate behavior to follow. For example, for mutual comprehension, the behavior is "take turns in conversation or ensure a balanced discussion."

Such principles may seem detached from many contexts in the rural South characterized by a highly unequal public sphere with strong social hierarchies and deeply entrenched social inequities. Inviting stakeholders to sit in one room and participate is not sufficient to enable inclusive governance. Fraser (1997) sees the public sphere idealized by Habermas (1984) as "an institutional mechanism for rationalizing political domination by rendering states accountable to (some of) the citizenry" (Fraser 1997, 72). It is important to recognize that deliberative governance goes beyond participation to involve negotiation (Leeuwis 2000). Furthermore, knowledge production and use are inherently political, especially when a high degree of complexity, uncertainty, and strong economic and political interests come into play (Forsyth 2003). Therefore, we cannot reduce deliberation as a mere exchange of information. Which knowledge is valued and who has authority matter as much as the content of the deliberative process.

We explored the use of deliberative principles to inform debates in the context of climate change governance in Nepal where scalar politics play an important role in shaping adaptation interventions (Yates 2012). Our analysis explored two main principles for assessing the quality of deliberation: (1) fairness and (2) competence; both commonly used to assess deliberation practices (Abelson et al. 2003; Webler 1995). Fairness refers to the equal distribution to act meaningfully, which we evaluated in terms of equal distribution of opportunities to speak freely among different groups of stakeholders and equal role within the deliberation process (Webler 1995). Competence, or effectiveness, refers to the extent to which new knowledge and understandings develop through the process (Webler 1995). For the analysis of competence, we also drew on the concept of discourse coalition (Hajer 1995) which is a grouping of actors around specific story lines when engaging in policy debates. We examined how actors positioned themselves around different story lines and engaged in new discourse coalitions.

3.3 Methodology

3.3.1 Research Area

The research took place in Dhanusa District, a district located in the plains of Tarai-Madhesh (Figure 3.1).

Located in the southeastern part of the Central Development Region, Dhanusa District covers an area of 1,188.7 km^2 with a population of 754,777 inhabitants (Central Bureau of Statistics 2012). The main cereal crops grown in the district are wheat and paddy. According to the National Sample Census of Agriculture (NSCA) (2011/2012), marginal farmers, operating landholding of less than 0.5 ha, represent 50 percent of landowners. Women own around 6 percent of the total landholding area (CBS 2013). As per the NSCA, 72 percent of the irrigable area is under irrigation with 27 percent of the irrigated area relying on groundwater (CBS 2013). However, considering the operation and maintenance issues that plague large public irrigation systems in the Tarai-Madhesh region (Roth and Vincent 2012), it is unclear to which extent the area irrigated under major and medium surface irrigation systems receives water in sufficient quantity and when needed. Furthermore, access to groundwater in the region is highly unequal due to monopolistic markets, lack of access to electricity, and high diesel prices that disadvantage tenant farmers (Sugden 2014).

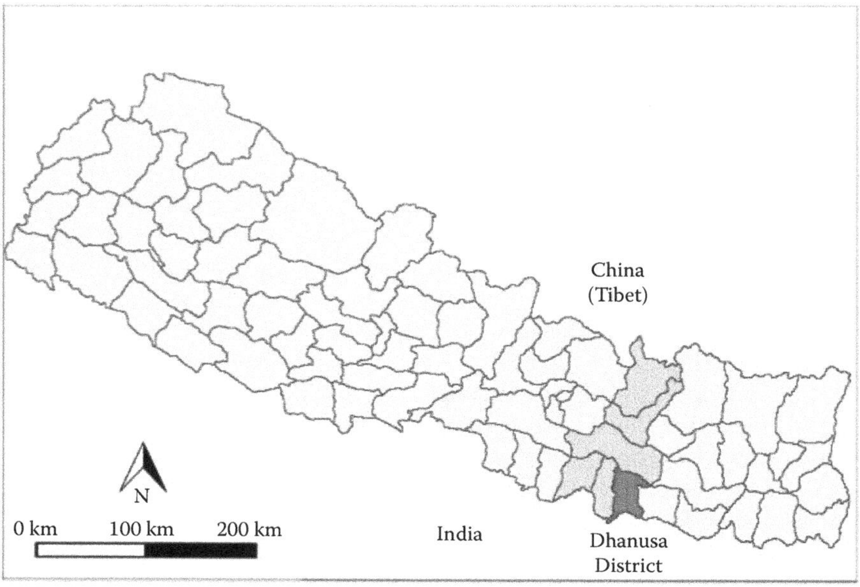

FIGURE 3.1
Location of Dhanusa District in the Central Development Region, Nepal.

Dhanusa District offered an interesting setting to conduct action research on vulnerabilities to climate change. It is prone to extreme climatic events (droughts and floods) and has one of the highest rates of male out-migration in the country, leading to dramatic social and economic changes over the last decade, notably in terms of gender roles and relationships, collective action, and monetization of the economy (Sugden et al. 2014). Male out-migration has resulted from a wide range of factors, such as increased monetization of the economy, increased price of agricultural inputs, and low access to markets, that go beyond climatic stress (Sugden et al. 2014).

3.3.2 Action Research

We developed the action research as a learning and iterative process that stretched over a period of 27 months between October 2012 and December 2014. Action research is a research process involving the concept of "changing-practice practice" (Kemmis 2009). Participatory action research was particularly well suited to address the research question as it allowed us not only to observe argumentation during the deliberative processes but also to link these to non-discursive practices through observations. The use of participatory video allowed us to capture farmers' own framings and story lines of their vulnerabilities with limited interference from researchers' framings.

After a pilot participatory video project conducted in November 2012, we initiated the research project with two groups of six men and women farmers from two village development committees[1] (VDCs) in Dhanusa District. By involving both men and women, our intention was to identify gendered differences in experiences and framing of vulnerabilities. The research team first conducted focus group discussions with men and women farmers separately about social and climate change, with the objective for the participants to identify and reflect upon the issues that they would like to develop in a film. Two film professionals then trained the two groups of farmers to use a video-camera, to frame questions, and to interview other farmers in their community. Farmers also learned how to do simple editing of their audio-visual material so that they could have control over the production process.

During the 10 months that followed, farmers conducted interviews of other farmers in their VDC on 12 topics of their choice. They edited the audio-visual material, which consisted of around 60 interviews, into 12 short films on climate change and societal change.[2] Journalists from the Nepal Forum of Environmental Journalists (NEFEJ) showed each film to two expert

[1] The Village Development Committee was formerly the lower administrative unit in Nepal. In Dhanusa District, the average population size is 7,400 persons in a VDC.

[2] Although our intention was that farmers have control over the overall production process, this proved in practice difficult as editing requires sitting for an extended period of time, and most farmers could not stick to that because of frequent power cuts and because of their busy schedule.

or policy maker representatives. The journalists from NEFEJ video-recorded the responses of the two representatives and edited them together with the farmers' film as 20-min television episodes. Each episode was broadcasted weekly on the national television under a new program called "Samudayako Aawaj" (in English, Communities' Voices) between July and December 2013. The last phase of the project aimed at initiating a deliberative process through the organization of a series of public dialogues between October and December 2014. These public dialogues used the audio-visual material from the 12 films compiled into a 30-min documentary to initiate the discussion.

3.3.3 Organization of the Public Dialogues

Various organizations in close collaboration with the research project leader organized the public dialogues, presented in chronological order in Table 3.1, according to different formats. These organizations were the Nepal Madhesh Foundation (NEMAF), a research and advocacy organization working on human rights and governance in the Tarai-Madhesh region;

TABLE 3.1

Characteristics of the Dialogues Organized

	Location	Format	Number of Participants for Each Type of Organization	Organized/ Facilitated by	Other
1	6 VDCs, Dhanusa District	Facilitated discussion based on a set of questions defined by the research team, in small groups of around 7 persons disaggregated by sex	Farmers: 185 (in total, for 6 dialogues) *Sex: 97 men, 88 women* *Caste/ethnicity: 3 Tarai Brahmin, 45 Dalits, 4 Janajatis, 6 Muslim, 127 Other Backward Caste*	NEMAF/NGO representative	
2	Janakpur, district headquarter, Dhanusa District	Facilitated discussion upon the discretion of the facilitator	Farmers: 17 Journalists: 2 District-level government officials: 4 *Sex: 14 men, 9 women*	Panos South Asia/A journalist	Recorded, edited and broadcasted on the radio
3	Janakpur, district headquarter, Dhanusa District	Facilitated discussion in small groups with representatives from each type of participant around 3 pre-defined themes	Farmers: 12 Civil Society organizations: 8 District-level government officials: 6 *Sex: 17 men, 9 women*	NEMAF/NGO representative	

(*Continued*)

TABLE 3.1 (*Continued*)

Characteristics of the Dialogues Organized

Location	Format	Number of Participants for Each Type of Organization	Organized/ Facilitated by	Other
4 Kathmandu	Facilitated discussion upon the discretion of the facilitator	Farmers: 10 Civil society organizations: 1 Media organizations: 2 National-level government agencies: 2 District-level government agencies: 2 *Sex: 10 men, 7 women*	Panos South Asia/A journalist	Recorded, edited and broadcasted on the radio
5 Kathmandu	Workshop style, facilitated discussion on relevant themes selected by the facilitator	Farmers: 10 Civil society organizations: 3 Politicians: 1 National-level government agencies: 2 District-level government agencies: 3 *Sex: 12 men, 7 women*	SIAS/A researcher	

PANOS South Asia, a media organization; and the South Asia Institute of Advanced Studies (SIAS), a research institute working on natural resource governance and climate change issues in Nepal.

The objective of these public dialogues was to engage a deliberation whereby the participants could discuss their views and perceptions of vulnerability, acknowledge and respect others' views, and engage in a constructive dialogue to elaborate a shared framing of the problem and to negotiate potential solutions.

3.3.4 Data Analysis

All the public dialogues were audio-recorded, transcribed, and translated in English. We coded and analyzed the transcripts using a qualitative data analysis software. We examine in this chapter three out of five events organized: the dialogues at the VDC level, one dialogue at the district level, and one final dialogue at the national level (dialogues 1, 3, and 4 in Table 3.2). The three events we selected are those we evaluated to be the closest to authentic deliberation as per the criteria of fairness and competence introduced earlier. We achieved fairness in these three events through a small group and break-up in small groups, an informal setting, and use of a facilitator who explicitly demonstrated through his tone, facilitation, and attitude that he was an impartial facilitator. In the last event (dialogue 4), some of the men

TABLE 3.2
Topics Selected by the Groups of Women and Men Farmers for Their Films

Film Number	Women Farmers	Men Farmers
1/2	Migration	Migration
3/4	Dowry	Changes in farming practices
5/6	Girls' futures	Collective action for irrigation
7/8	The failure of agriculture	How to reduce migration
9/10	Changes in gender norms	Roads
11/12	Roads	Education

and women farmers took turns filming the event with their video-camera, thereby modifying positions of power in different ways: by standing outside the tables above other participants, by demonstrating their capacity to use a technology, and by moving from the position of a participant to that of an observer and producer of knowledge. We evaluated competence through the analysis of the transcripts by comparing framings of vulnerabilities and the solutions during the event with farmers' framing in their films and the framing of vulnerabilities in climate change policies.

3.3.5 Limitations

The analysis we presented focused mostly on shifts in discourses and we did not elaborate in detail how identities and hierarchies also affected deliberative practices. The main limitations of the action research process lie in the limited number of events that we organized. Farmers participated in three events within 3 month, whereas most other stakeholders only participated in one event. As several scholars note, the value of deliberation is not only in the reasoning, learning, and reframing process but also in the social relationships that develop through the regular interaction of participants. If not necessarily leading to more democratic decisions, deliberative practices can provide a useful basis to restore trust (Bloomfield et al. 2001; Innes and Booher 2003). We regret as well that we had no opportunity to follow-up on the solutions that the participants proposed.

3.4 Contrasting Framings of Vulnerability

3.4.1 Climate Change Policies in Nepal

Nepal's political involvement in climate change started in 1992 when it became a party to the UNFCCC. The NAPA (Government of Nepal (GoN) 2010) was Nepal's first comprehensive public policy document on climate.

GoN developed it through a largely elite-dominated technocratic process, despite its claims of being inclusive and participatory, and gave rise to little debate or contestation (Ojha et al. 2016). It was important as it prioritized adaptation actions, paving the way for the design and implementation of nine climate change adaptation projects designed with a total budget of 350 million U.S. dollars (Sova and Chaudhury 2013).

The NAPA typically adopts a risk-hazard approach to climate change adaptation (Figure 3.2) whereby the analysis starts from climatic hazards and traces their potential impacts on a range of sectors or domains to propose adaptation options to address these impacts (Ribot 2010).

The UNFCCC designed the adaptation responses proposed in the NAPA to address climatic risks and they are mostly technical and managerial options (e.g., construction of water storage, adoption of drought-resistant crop varieties, and organic farming practices) defined for each sector or domain in isolation. In practice, climate change interventions have indeed failed to challenge structural causes of vulnerability (Nagoda 2015). This type of framing is not specific to Nepal but is in line with dominant international discourses on climate change that frame climate change as a technical problem (Dewulf 2013).

Nepal developed the Climate Change Policy (GoN 2011) to comply with the institutional developments required of all signatory parties by the UNFCCC to promote climate adaptation and mitigation and mobilize the financial resources necessary to climate change adaptation programs and to

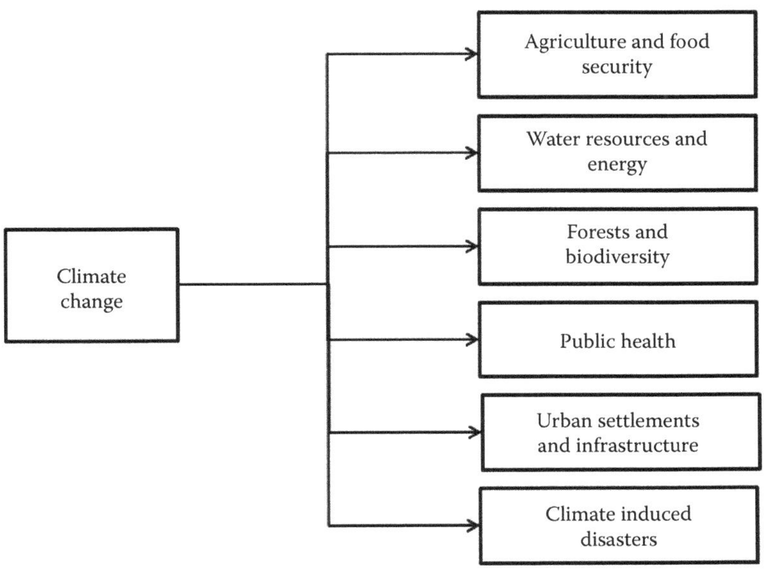

FIGURE 3.2
Analytical framework guiding Nepal's NAPA. (GoN 2010.)

implement these programs according to the national development agenda. The emphasis is also on climate change impacts on socio-economic sectors, for example, "*National efforts to make the socio-economic sectors climate-resilient*" (GoN 2011, Section 4.2). In line with the NAPA, Nepal's Climate Change Policy clearly positions technologies as the most promising tool for adaptation. The proposed adaptation responses are technico-managerial and focus on capacity building.

Although Nepal was one of the last LDCs to embark on a NAPA process, it remarkably extended the process to develop a national framework on Local Adaptation Plans for Action (LAPAs). The objective of the LAPAs is to ensure the integration of adaptation and resilience into local and national planning processes and as well to meet the requirement of the Climate Change Policy that 80 percent of the funds flow to local institutions and communities (GoN 2011). Nepal has developed and implemented around 100 LAPAs in 13 districts in Mid and Far West Nepal under the National Climate Change Support Program, a multi-donor initiative.

3.4.2 Framings of Vulnerabilities in Farmers' Films

Before examining farmers' framings of vulnerabilities in the films, it is interesting to note that none of the topics chosen by farmers for their films specifically focused on climate change (Table 3.2). The first topic selected by both the group of men and women farmers was on migration—a phenomenon affecting not only daily lives but also, as shown in the films, farming, local collective action, access to irrigation water, and monetization of the dowry system.

In the farmers' films, the central component of their vulnerability was "the failure of agriculture"—an issue that does not only impact the economic and material aspect of their lives but also touches upon farmers' identities and has far-reaching repercussions on social relationships, institutions, and farming practices. The failure of agriculture has been one of the major driving factors of long-term male out-migration in India and overseas, creating new forms of vulnerabilities for women left behind through increased workload and new responsibilities (Sugden et al. 2014). Male out-migration in the Tarai region has also increased land commodification, patterns of land use, and tenancy relations (Sunam and McCarthy 2015).

The issue of increased climatic variability and unpredictability is only one of the factors that contribute to this failure (Figure 3.3). Similarly, a "development first approach" (Davidson et al. 2003) which acknowledges that a wide range of non-climatic factors affect vulnerability (Nightingale 2015; O'Brien and Leichenko 2000; Ribot 2010). The development's first approach sharply contrasts with the risk-hazard perspective that underpins national climate policy documents in Nepal. National policies consider climatic risk as the point of departure for analysis and action and link it to different sectors through a linear causal relationship. As different interpretations

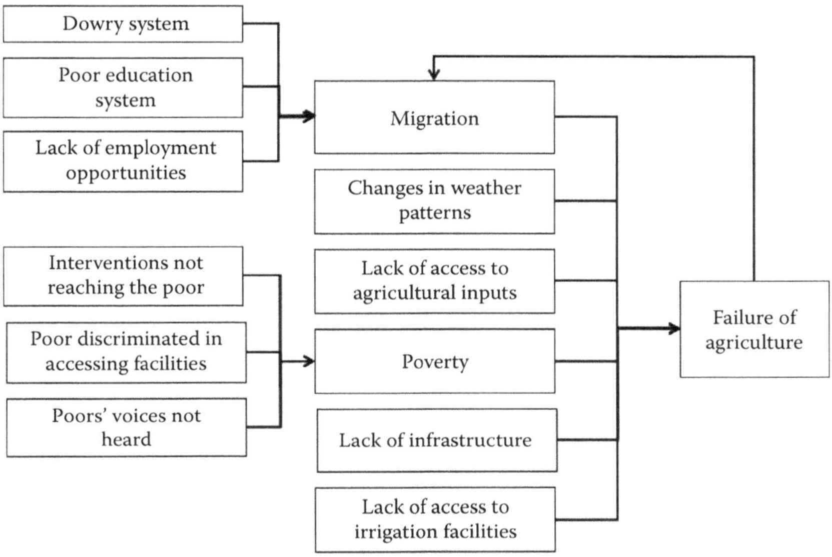

FIGURE 3.3
Farmers' causal chain of vulnerability. (*Source*: Created by the authors from farmers' film.)

of vulnerabilities are rooted in different sets of discourses and knowledge, the reliance of one over another greatly matters in the way countries design climate change adaptation policies and in the choice of adaptation activities (O'Brien et al. 2007).

In their films, farmers report their observations of change in rainfall patterns compared to 20 years ago, with less frequent and more irregular rainfall events now. But they clearly articulate that this is a problem only because they lack entitlements to irrigation. They state that there is no water in public canals because *"the river is dammed upstream and they [the government] do not allow water to come downstream"* (male farmer, Film 6) and blame the government for neither providing tube wells nor electricity for pumping. They also stress the lack of response they received from district-line government agencies and local government bodies when they filled in applications to get funding to build a concrete dam for irrigation.

Farmers refer to proximate causes of vulnerability such as lack of capacity, lack of employment, poverty, and lack of infrastructures but they systematically link these to structural factors. For example, if poverty is a cause of vulnerability, it is not only because poor people lack money to buy services, commodities, and access facilities, but also because institutional and political structures discriminate against the poor. They point to a defunct government system which does not adequately consider the voices of the poor people, raising issues of distributive justice and representation: *"There is no electricity for irrigation…uh… Nothing, I mean the government does not provide these facilities. Only the rich people receive benefits, and nobody listens to the poor"* (female farmer, Film 4) and "*We raised this issue. We also requested the secretary and technician of the VDC, but nobody addressed our concern regarding roads. This happens nationwide. We are poor so nobody cares"* (male farmer, Film 9). They also highlight the existence of competing parallel private and public systems in which poor people, or more broadly common people (those who are not politically connected), are not able to afford the private services that should be accessible to everyone through public agencies, such as good quality education: "*Teachers in public schools educate their own children... uh... in private schools. Public schools are only the choice of poor children"* (male farmer, Film 11). Lastly, women in their films also stressed their low legitimacy to participate in and influence public decision-making.

Farmers were very critical of the government in their films: for them, the lack of effectiveness and fairness in the provision of government services is the main cause of the failure of agriculture and of their vulnerability. What farmers coined as "the failure of agriculture" goes beyond crop failure. It has resulted from a combination of factors from increased wages for laborers and labour shortage due to male out-migration, to lack of irrigation, market access, and increased costs of inputs and diesel. Farmers clearly used the videos to put forward their claims and requests to the government. They explicitly asked for infrastructures and irrigation facilities (e.g., dams, roads, borewells) and improved access to agricultural inputs and extension services. There is a recurrent motto in farmers' films that the government has neglected and abandoned farmers. Several farmers interviewed in the films refer to the contradiction between Nepal's core identity of being an agricultural country and the blatant neglect of agriculture and farmers by the government that farmers feel in the region.

In summary, the films revealed a profound disconnect between farmers' framings and experiences of vulnerabilities and how national climate change adaptation policies and programs framed vulnerability. The films also provided evidence concerning the diversity of experiences of vulnerability across farmers, for example, depending on gender and age. Lastly, the films revealed a high level of distrust between farmers and the government, highlighting potential difficulties to initiate an inclusive, fair, and effective deliberative process.

3.5 A Step Towards Deliberative Governance

3.5.1 VDC Level: Farmers Only

The first dialogues organized in six VDCs in Dhanusa District were for farmers. The objective was to trigger a reflection on vulnerabilities among farmers in a non-threatening environment and, at the same time, collect data that could inform the next round of dialogues involving government and civil society representatives. The dialogues gathered in total 97 men and 88 women from different castes, ethnicities, and religions, with 69 percent of farmers from other backward castes, 24 percent from Tarai, and 7 percent Muslim, Tarai Brahmin, and Tarai Janajatis. Because we mixed the discussion groups, we could not explore the difference of framings between identities except for gender.

The farmers who participated in the dialogues stressed their limited agency and how structural factors have constrained the livelihood choices available to them. Specifically, they highlighted their low capacity to maintain and develop productive and profitable farming systems. Many women also voiced their inability to pursue the life choices they wanted to and expressed their hope that their children will have better opportunities through education. Both men and women felt that the government knew about the problems related to migration, farming, irrigation education, and so on, raised in the films. They did not think that the lack of response from the government was a lack of knowledge but rather due to a corrupt and dysfunctional government system, characterized by low accountability and lack of representation. In contrast to the films, farmers also engaged in a self-reflection and in several dialogues, both men and women farmers repeatedly pointed out the lack of social cohesion and cooperation within their community across dialogues. They felt the latter was a barrier to conduct collective action and hold the government accountable. They attributed this lack of social cohesion to male out-migration. According to them, the lack of public action has resulted from both male out-migration and a fixed division of gender roles which limits women's involvement in the public sphere.

Despite their limited capacity to influence policies and improve the governance system, male and female farmers proposed many actions where they would play a role to make line agencies or public schoolteachers more accountable to them, thus moving beyond being passive victims and beneficiaries to becoming active agents of social change. For example, they proposed to put pressure on the government to take necessary measure to control the black market in agricultural inputs. They also proposed to develop better relationships between schoolteachers and parents to increase teachers' sense of responsibility and duty in providing quality education.

3.5.2 District Level: Farmers and Government Line Agencies

We invited two farmers from the group of farmers' participants from each of the six VDCs to the district-level dialogue. This event also included government officials from district-line agencies working on agriculture, education, women's development, and water-induced disasters. After the screening of the film, we formed three groups around three topics: education, roads, and agriculture. Each group consisted of four farmers and one or two government officials. The government officials who participated in the dialogue were middle-class officials who frequently interact with farmers in their function. The hierarchy between them and the farmers was thus not great which led to rather open and informal discussions that we noted from observations of facial cues, body positions, and gestures. The groups first had to identify the main issues related to their topic whether outlined in the film or not. The objective was for the participants to realize that there are different framings of the issues related to each of these topics. After the initial discussion to identify important issues, each group discussed potential solutions for development and implementation across various institutional levels. Each group had to nominate a representative to present the results of their discussion. The farmers and government officials groups selected both men and women to present.

The group on the topic of roads interestingly came up with a model of local governance that touched upon issues of representation, inclusion, and justice:

- Primarily, they will set up committees comprising the civil society and government employees at the local level for dialogue and discussion on development activities.
- While forming such committees, they will encourage female participation and they will provision that a woman may hold a post as a chairperson of the committees.
- While planning, they will call upon every representative from every community or group for discussions.
- In case of irregularities, they will report it to concerned agencies and they will call upon proper and fair investigation.

Such proposals largely go beyond the requests made by farmers in their films for more infrastructure development and provide response to the farmers' identification of political economic structures as root causes of vulnerabilities, which national climate change adaptation policies had largely ignored. The proposals' intent is to institutionalize deliberation among different parties, restoring trust and dialogue while also addressing issues related to male out-migration. Such deliberation platforms between farmers and government officials have been specifically missing because of the lack of locally elected government representatives for the last 15 years in Nepal. Until recently, this proposal relates to issues of procedural justice and outlines that farmers are not just expecting fairer service provision (distributive justice) but also the right to participate in and influence local development decisions and to support bottom-up planning.

The proposal of a provision allowing a woman to be the chairperson is also innovative in Nepal. So far, the national regulations have indicated a minimum of 33 percent female participation in grassroots organizations without any mention of the type of position. In practice, most of the chairpersons for all influential groups (water user and community forest user groups) are male. Reserving the chairperson position for women can in practice become tokenistic, yet it is remarkable that the propositions go beyond current public policies by specifically reserving the most important seat to women. Lastly, the solution proposed even includes a sanction mechanism in case of deviation from the proposed institutional arrangement, with clear responsibilities shared between the citizens (by reporting to concerned agencies) and public authorities (by calling for a proper and fair investigation).

3.5.3 National Dialogue: Farmers, Civil Society, and Government Officials

During the national dialogue, the farmers invited were the group of men and women farmers who had directed the films. They had already taken part in one dialogue at the district level with government line agencies and representatives from district level government bodies (number 2 in Table 3.2, which we do not discuss in this chapter), so they already had prior experience with deliberation. We also invited some of the district government officials who had already taken part to the dialogue in Dhanusa District with farmers.

Although it is not the purpose of this chapter to elaborate on the power and discourses interplay during the national level dialogues, there were interesting indications of changes in discourse coalitions during this event, with some government officials positioning themselves along story lines that strikingly differed from the technico-managerial narratives dominating national policies. For example, one of the district officials related the lack of access to water raised by farmers to both the lack of cooperation among villagers and the disastrous effects of local politics within a dysfunctional governance system:

> [...] *now people depend more on the government. Now in villages there is no feeling of cooperation. The political system has entered down to the village level and destroyed the feeling of cooperation.* [...] *we can't do even simple work like constructing dams due to them [political parties]. If a political party takes steps, another will oppose* [...]. *A woman participant just mentioned about a water channel destroyed in her village. You cannot build this channel again because the politics will interfere. So, we have to solve these factors too.*

This type of story line clearly differs from the technocratic risk-hazard perspective that dominates the NAPA and Climate Change Policy. As Hajer (1995) argues, individuals do not stick to one set of discourses but can join a different discourse coalition depending on the settings. These possibilities can be opportunities for legitimizing new knowledge claims and triggering institutional change.

The solutions put forward during the national level dialogue were in line with the story lines present in farmers' films (e.g., creating job opportunities to reduce out-migration will reduce farmers' vulnerabilities in the region) and dominant government and international climate change adaptation story lines (e.g., building farmers' capacity to adopt good farming practices and water-smart irrigation techniques is key to reduce vulnerability). However, a work meeting gathering the journalist who was facilitating the event, a politician former minister, a government official, and two project partners (and the author, project leader and principal investigator, and the media partner) followed the national level dialogue. During this meeting, the journalist, who is the editor of a well-known English language weekly newspaper in Nepal, indicated that the dialogue had made him realize that farmers' vulnerabilities and agricultural issues were invisible in the national media and specifically in the press. In his view, this was because the media's bias towards Kathmandu, the centre of power, and that the media does not perceive agriculture as a glamorous topic. Journalists prefer to write about national politics, a topic that better serves their professional growth and career. The small group decided on a few action points including training young business journalists to increase the coverage of issues related to agriculture and farmers; introducing columns for farmers and encouraging journalists to write more about agriculture; making the agricultural sector more glamorous through the use of social media; and publishing a magazine dedicated to farmers and their work. The dialogue therefore brought forward critical issues of recognition that were at the core of farmers' claims: the lack of public attention to farmers in a country whose identity and gross domestic product are agricultural.

3.6 Discussion and Conclusion

Scholars have pointed to the tension between deliberative procedure and the outcomes of deliberation (Niemeyer and Dryzek 2007). Our findings indicate that, even within an unequal public sphere, deliberation can support the opening of discourses towards the development of shared understandings of farmers' multiple experiences of vulnerabilities. Specifically, discourse analysis evidenced the emergence of new story lines during the dialogues—story lines acknowledging the structural causes of vulnerability that are absent from national climate change policies. We also observed a repositioning of a few government officials around different discourse coalitions during the dialogues that took place. These findings are encouraging in the sense that they imply that deliberative governance could support institutional reforms that represent a broader set of perspectives thereby opening climate change policies and discourses beyond a risk-hazard perspective and techno-managerial solutions. The deliberations did not lead to a meta-consensus (Niemeyer and Dryzek 2007) on the complex and multi-scale causes of vulnerabilities. Yet the broadening from discourses in counter-narratives to dominant story lines is a promising first step. We also observed that deliberation initiated self-reflection notably among farmers and some of the media actors participating in the dialogue.

However, as we know, fostering deliberative governance requires more than inviting stakeholders to sit together. The two other events that we organized, and which we did not discuss in this chapter, were less successful in bringing a real dialogue between farmers and government officials. The facilitation proved to be essential in framing the terms and scope of the debates. In one dialogue, the facilitator inadvertently created a discursive closure which kept the discussion between farmers and government officials very polarized. The preparation and reflection of farmers prior to the dialogue also made a difference between the two district level workshops that we organized. Lastly, although we did not elaborate on it in this chapter, the social hierarchies among participants were also critical (Clement 2018). We posit that a realistic and meaningful objective for deliberation is for the participants to acknowledge the existence of multiple framings and specifically support the emergence of counter-narratives to hegemonic discourses. This objective implies that researchers must identify and unpack these framings and different story lines to guide governance processes on what "is" rather than providing top-down arguments on what should be.

Acknowledgments

The Climate Change, Agriculture and Food Security (CCAFS), and Water, Land and Ecosystems (WLE) CGIAR programs funded this research.

References

Abelson, J., P.-G. Forest, J. Eyles, P. Smith, E. Martin, and F.-P. Gauuvin. 2003. Deliberations about deliberative methods: Issues in the design and evaluation of public participatory processes. *Social Science & Medicine* 57:239–251.

Arndt, C., and F. Tarp. 2017. Aid, environment and climate change. *Review of Development Economics* 21:285–303.

Ayers, J. 2011. Resolving the adaptation paradox: Exploring the potential for deliberative adaptation policy-making in Bangladesh. *Global Environmental Politics* 11(1):62–89.

Bloomfield, D., K. Collins, C. Fry, and R. Munton. 2001. Deliberation and inclusion: Vehicles for increasing trust in UK public governance? *Environment and Planning C: Government and Policy* 19:501–513.

Byg, A., and J. Salick. 2009. Local perspectives on a global phenomenon—Climate change in Eastern Tibetan villages. *Global Environmental Change* 19(2):156–166.

Central Bureau of Statistics. 2012. *National Population and Housing Census 2011* (Village Development Committee/Municipality). Kathmandu, Nepal: Government of Nepal, National Planning Commission.

Central Bureau of Statistics. 2013. *National Sample Census of Agriculture Nepal 2011/12*. Central Bureau of Statistics, Kathmandu, Nepal: Government of Nepal, National Planning Commission Secretariat.

Clement, F. 2013. For critical social-ecological system studies: Integrating power and discourses to move beyond the right institutional fit. *Environmental Conservation* 40(1):1–4. doi:10.1017/S0376892912000276.

Clement, F. 2018. Exploring the use of audiovisual media for deliberative dialogues: Reframing discourses on vulnerabilities to climate change in Nepal. In *Action Research in Policy Analysis: Critical and Relational Approaches to Sustainability Transitions* (Eds.) K. Bartels and J. M. Wittmayer, Chapter 10. Abingdon, UK: Routledge.

Davidson, O., K. Halsnæs, S. Huq, M. Kok, B. Metz, Y. Sokona, and J. Verhagen. 2003. The development and climate nexus: The case of sub-Saharan Africa. *Climate Policy* 3(Supplement 1):S97–S113.

Dewulf, A. 2013. Contrasting frames in policy debates on climate change adaptation. *Wiley Interdisciplinary Reviews: Climate Change* 4(4):321–330. doi:10.1002/wcc.227.

Dore, J. 2014. An agenda for deliberative water governance arenas in the Mekong. *Water Policy* 16:194–214.

Dryzek, J. S. 2000. *Deliberative Democracy and Beyond*. Oxford, UK: Oxford University Press.

Eriksen, S., P. Aldunce, C. S. Bahinipati, R. D. Martins, J. I. Molefe, C. Nhemachena, K. O'Brien et al. 2011. When not every response to climate change is a good one: Identifying principles for sustainable adaptation. *Climate and Development* 3(1):7–20.

Eriksen, S., A. J. Nightingale, and H. Eakin. 2015. Reframing adaptation: The political nature of climate change adaptation. *Global Environmental Change* 35:523–533.

Fischer, F. 2003. *Reframing Public Policy: Discursive Politics and Deliberative Practices*. Oxford, UK: Oxford University Press.

Forsyth, T. 2003. *Critical Political Ecology: The Politics of Environmental Science*. New York: Routledge.

Forsyth, T. 2011. Expertise needs transparency not blind trust: A deliberative approach to integrating science and social participation. *Critical Policy Studies* 5(3):317–322.

Foucault, M. 1975. *Surveiller et Punir. Naissance de la Prison*. Paris, France: Gallimard.

Fraser, N. 1997. *Justice Interruptus: Critical Reflections on the "Postsocialist" Condition*. New York: Routledge.

Gastil, J., and L. W. Black. 2008. Public deliberation as the organizing principle of political communication research. *Public Deliberation* 4(1):3.

GoN. 2010. *National Adaptation Programme of Action (NAPA) to Climate Change*. Kathmandu: Government of Nepal (GoN), Ministry of Environment.

GoN. 2011. *Climate Change Policy*. Kathmandu: Government of Nepal (GoN), Ministry of Environment.

Habermas, J. 1984. *The Theory of Communicative Action. Volume 1: Reason and the Rationalization of Society*. London, UK: Heinemann.

Hajer, M. J. 1995. *The Politics of Environmental Discourse: Ecological Modernization and the Policy Process*. Oxford, UK: Oxford University Press.

Hajer, M., and H. Wagenaar, eds. 2003. *Deliberative Policy Analysis: Understanding Governance in the Network Society*. Cambridge, UK: Cambridge University Press.

Innes, J. E., and D. E. Booher. 2003. Collaborative policy-making: Governance through dialogue. In *Deliberative Policy Analysis: Understanding Governance in the Network Society* (Eds.) M. Hajer and H. Wagenaar, pp. 33–59. Cambridge, UK: Cambridge University Press.

Jones, L., and E. Boyd. 2011. Exploring social barriers to adaptation: Insights from Western Nepal. *Global Environmental Change* 21(4):1262–1274.

Kemmis, S. 2009. Action research as a practice-based practice. *Educational Action Research* 17(3):463–474.

Leach, M., G. Bloom, A. Ely, P. Nightingale, I. Scoones, E. Shah, and A. Smith. 2007. Understanding governance; pathways to sustainability. *STEPS Working Paper 2*. Sussex, UK: IDS.

Lebel, L. 2013. Local knowledge and adaptation to climate change in natural resource-based societies of the Asia-Pacific. *Mitigation and Adaptation Strategies for Global Change* 18(7):1057–1076.

Leeuwis, C. 2000. Reconceptualising participation for sustainable rural development: Towards a negotiation approach. *Development and Change* 31:931–959.

Merino, E., and J. C. Ribot. 2012. Special issue introduction: Adding insult to injury: Climate change and the inequities of climate intervention. *Global Environmental Change* 22(2):323–328.

Methmann, C., D. Rothe, and B. Stephan. 2015. *Interpretive Approaches to Global Climate Governance: (De)constructing the Greenhouse*. New York: Routledge.

Nagoda, S. 2015. New discourses but same old development approaches? Climate change adaptation policies, chronic food insecurity and development interventions in northwestern Nepal. *Global Environmental Change* 35:570–579.

National Adaptation Programme of Action (NAPA) to Climate change. 2010. *Climate Change Policy*. 2011.

Neef, A. 2009. Transforming rural water governance: Towards deliberative and polycentric models? *Water Alternatives* 2(1):53–60.

Niemeyer, S., and J. S. Dryzek. 2007. The ends of deliberation: Meta-consensus and inter-subjective rationality as ideal outcomes. *Swiss Political Science Review* 13(4):497–526.

Nightingale, A. J. 2015. A socionature approach to adaptation: Political transition, intersectionality, and climate change programmes in Nepal. In *Climate Change Adaptation and Development: Transforming Paradigms and Practices* (Eds.) H. Inderberg, S. Eriksen, K. O'Brien, and L. Sygna, pp. 219–234. New York: Routledge.

O'Brien, K., S. Eriksen, L. P. Nygaard, and A. N. E. Schjolden. 2007. Why different interpretations of vulnerability matter in climate change discourses. *Climate Policy* 7(1):73–88.

O'Brien, K., and R. Leichenko. 2000. Environmental change and globalization: Double exposure. *Global Environmental Change* 10:221–232.

Ojha, H., A. Pain, S. Ghimire, A. J. Nightingale, D. B. Khatri, and H. Dhungana. 2016. Policy without politics: Technocratic control of climate change adaptation policy making in Nepal. *Climate Policy* 16(4):415–433.

Ribot, J. C. 2010. Vulnerability does not just fall from the sky: Toward multi-scale pro-poor climate policy. In *Social Dimensions of Climate Change: Equity and Vulnerability in a Warming World* (Eds.) R. Mearns and A. Norton, pp. 47–75. Washington, DC: The World Bank.

Roe, E. 1991. Development narratives, or making the best of blueprint development. *World Development* 19(4):287–300.

Roth, D., and L. Vincent. 2012. Irrigation technology and irrigation management reform: The case of the Terai Region in Nepal. In *Controlling the Water: Matching Technology and Institutions in Irrigation Management in India and Nepal* (Eds.) D. Roth and L. Vincent, pp. 112–141. Oxford, UK: Oxford University Press.

Smucker, T. A., B. Wisner, A. Mascarenhas, P. Munishi, E. E. Wangui, G. Sinha, D. Weiner, C. Bewenge, and E. Lovell. 2015. Differentiated livelihoods, local institutions, and the adaptation imperative: Assessing climate change adaptation policy in Tanzania. *Geoforum* 59:39–50.

Sova, C., and A. Chaudhury. 2013. *State of Agricultural Climate Change Adaptation Policy in Nepal*. Copenhagen, Denmark: Retrieved from Copenhagen.

Stevenson, H., and J. S. Dryzek. 2015. *Democratizing Global Climate Governance*. Cambridge, UK: Cambridge University Press.

Sugden, F. 2014. *Landlordism, Tenants and the Groundwater Sector: Lessons from Tarai-Madhesh*. IWMI Research Report 162. Colombo, Sri Lanka: IWMI.

Sugden, F., N. Maskey, F. Clement, V. Ramesh, A. Philip, and A. Rai. 2014. Agrarian stress and climate change in the Eastern Gangetic Plains: Gendered vulnerability in a stratified social formation. *Global Environmental Change* 29:258–269.

Sunam, R. K., and J. F. McCarthy. 2016. Reconsidering the links between poverty, international labour migration, and agrarian change: critical insights from Nepal. *The Journal of Peasant Studies* 43(1):39–63.

Webler, T. 1995. 'Right' discourse in citizen participation: An evaluative yardstick. In *Fairness and Competence in Citizen Participation: Evaluating Models for Environmental Discourse* (Eds.) O. Renn, T. Webler, and P. Wiedemann, pp. 35–86. London, UK: Kluwer.

Yates, J. S. 2012. Uneven interventions and the scalar politics of governing livelihood adaptation in rural Nepal. *Global Environmental Change* 22(2):537–546.

Zachrisson, A. 2010. Deliberative democracy and co-management of natural resources: Snowmobile regulation in western Sweden. *International Journal of the Commons* 4(1):273–292.

4

Waste or Savior? Two Cases of Emerging Wastewater Irrigation in Urbanizing Kathmandu Valley

Silvia Quarta, Dik Roth, Robert Dongol, Anushiya Shrestha, and Saroj Yakami

CONTENTS

4.1 Introduction 51
4.2 Research Approach, Methods, and Locations 53
4.3 Development of Wastewater Irrigation in Kathmandu Valley 56
4.4 Two Cases of Wastewater Irrigation 58
  4.4.1 Bhaktapur 58
    4.4.1.1 History of a Sewage Pipe 58
    4.4.1.2 Golmadi: Wastewater as an Opportunity 61
  4.4.2 Dadhikot 61
    4.4.2.1 New Irrigation Practices, Technologies, and Problems 63
    4.4.2.2 Dadhikot: Growing Pressures on Irrigated Agriculture 64
4.5 Analysis: Multiple Agencies, Assemblages, and Changing Access to Water 65
4.6 Conclusion 67
References 68

4.1 Introduction

Nepal is an extremely water-rich country, but at the same time suffering from water stress due to increasingly intensive water use, high seasonal rainfall variability, and the growing impact of climate change (Shrestha and Sada 2013; Shrestha et al. 2014). In Kathmandu Valley, the most urbanized and most rapidly urbanizing area of Nepal and the location of the capital Kathmandu, water stress levels are rising. Ninety percent of its 1,868 mm of annual rainfall is between June and September (Raschid-Sally et al. 2008). The main river in the valley, Bagmati, is subject to seasonal fluctuations.

Its basin collects 2 percent of the surface water available in the country but serves 8 percent of Nepal's population, causing a potential mismatch of supply and demand (WECS 2011). Annual groundwater consumption in the valley is now exceeding twice the amount of groundwater recharge (Shukla et al. 2011).

As a result of population increase in the valley, rapid changes in land and water use are taking place both in more urbanized and in peri-urban areas. Concurring with other authors on peri-urbanity (e.g., Iaquinta and Drescher 2000; for South Asia, see Narain 2009, 2014; Narain and Prakash 2016), we approach the peri-urban in terms of processes and flows rather than fixed and clearly bounded and governed spaces. What are often referred to as peri-urban areas are actually highly dynamic, evolving, constantly expanding, and shifting intermediate zones between cities and rural areas that are better understood in their processual dynamics—in terms of interconnections and flows of people, natural resources, goods and services, knowledge, capital—than as fixed states (Iaquinta and Drescher 2000, 18). Typically, peri-urban areas have both urban and rural characteristics, visible in a patchwork landscape of housing plots and high-rise buildings, agricultural fields, industries, and other land uses.

In this chapter, we explore these interconnections and flows between the urban and the rural (Narain 2016). Our focus in these explorations is the changing water uses and the growing role of polluted or wastewater for irrigation in these hybrid, peri-urban waterscapes (Swyngedouw 1999), which are increasingly also wastewaterscapes (Karpouzoglou and Zimmer 2016). In Kathmandu Valley, they manipulate and divert myriad flows of water, then access and use the water for production, consumption, and disposal of waste. Water flows interconnect the urban, peri-urban, and rural through rivers and canals, sewers, bottles, and water tankers, conveying surface and groundwater, fresh or polluted with waste and other forms of contamination, but also crucial for drinking and other domestic uses, and for agriculture, brick-making, and other purposes.

These processes are not neutral: they may create new opportunities and access for some but insecurity and exclusion for others, change the realities of water rights and access to water, and lead to conflicts between competing uses, users, and claimants of water (Swyngedouw et al. 2002). Poor and marginal groups with little bargaining power but highly dependent on land and water for their livelihoods, such as peri-urban farmers, tend to be at the losing end (Narain et al. 2013; Narain and Prakash 2016). With the economy of Nepal largely based on agriculture (Bartlett et al. 2010; Water and Energy Commission Secretariat 2011), the primary livelihood source for more than two-third of the population (Dhakal 2011), changes in access to land and water can deeply influence the lives of peri-urban farmers who may experience greater water insecurity.

Little is known of how farmers experience and perceive such changes nor how they cope with them in a materially, socially, and institutionally complex and changing environment (Iaquinta and Drescher 2000; Narain and Prakash 2016). This knowledge is also lacking for irrigation practices in peri-urban environments. Many studies focus on wastewater irrigation practices, related water quality and health issues, and farmers' awareness and perceptions of these from a more normative perspective (Rutkowski et al. 2007; de Leeuw 2014). This chapter, on the contrary, aims to develop a different understanding of these changing water uses. Rather than normatively labeling emerging irrigation practices as either good or bad or gathering evidence on water quality, we intend to bring forward the perceptions, motives, and choices (or lack of them) behind these practices as they are developing in specific water use contexts of Kathmandu Valley. This approach can help us gain a better understanding of these changes and can point us to new ways of dealing with peri-urban inequalities in access and water insecurities in the future.

We structured this chapter as follows. After this introduction, we present the research approach, methods, and case study locations. We then give a short overview of the development of wastewater irrigation in Kathmandu Valley. In the fourth section, we present the two cases: first, the Bhaktapur case (wastewater irrigation from a sewage system); and, second, the Dadhikot case (the pumping of highly polluted water from Hanumante river). After an analysis of the cases we provide the conclusion to this chapter.

4.2 Research Approach, Methods, and Locations

Political ecology of water and urban political ecology of water, interdisciplinary and socio-technical approaches in irrigation studies, actor-network and assemblage theory, and the anthropology of law conceptually and theoretically inspired the research for this chapter. The political ecology of water contributed important ideas about the hybrid, co-constituted character of nature, technology, and society, approaching them as socio-natures, hydro-social, and social-environmental networks, relationships, and processes (Swyngedouw 1999; Budds 2009). The urban political ecology of water (Swyngedouw et al. 2002) stresses the political, power-laden, and contradictory character of these processes. Seeing urbanization as a "historically specific accumulation of socio-environmental processes as well as the arena through which these transformations take place" (Swyngedouw 2002, 126), Swyngedouw notes that these transformations are never socially or environmentally neutral. Power relations shape them and they shape power relations that create differences in access and control, and inclusion and exclusion.

Actor-oriented sociology further inspired our research (Long 2001). Related, but moving beyond human agency, is Actor-Network Theory (ANT), which also considers the agency of material and non-human "actants," thus transcending ontological differences between people and objects (Law 1992; Latour 1999). In ANT, the social refers to "a type of connection between things that are not themselves social" (Latour 2005, 5), and society becomes a heterogeneous network because various materials are social agents interacting with each other (Law 1992). ANT thus provides a valuable tool for analysis, transcending the separation between irrigation technology and the practices and organizational arrangements around it. In peri-urban networks, multiple human actors with conflicting interests, values, framings of reality, and strategies interconnect with resources, technologies, institutions, laws, policies, and discourses and interventions of development. The fluid and scalar nature of these connections, the "labour of assembling and re-assembling socio-material practices that are diffuse, tangled and contingent" (Anderson and McFarlane 2011, 125), the socio-spatial and socio-material transformations that characterize peri-urban irrigation, and issues of agency in tempo-spatial orders (Li 2007; Allen 2011) also point to assemblage theory (Anderson and McFarlane 2011). Finally, to better understand what urbanization and changes in water flows mean for changing practices of claiming and accessing water, we used elements of property rights and access theories (Ribot and Peluso 2003; von Benda-Beckmann et al. 2006).

This chapter is based on fieldwork carried out by the first author of this chapter, Silvia Quarta, in Kathmandu Valley in 2016.[1] In view of the aims of this chapter, the authors designed the research as an ethnographic case study. Using ethnographic approaches yields the most comprehensive and in-depth understanding of the selected cases. Case studies can provide answers to questions like: Why do irrigators change from freshwater into wastewater sources? What does this mean for their water use and organizing practices? What agency do they exert in the complex emerging assemblages of resources, technologies and managing institutions? Case study approaches help to situate researched issues in a specific locale, explore linkages with other issues, and make sense of the emergent assemblages of land uses, water flows, human actors, organizational practices, technologies, and regulations. Providing a context-dependent understanding of processes, relationships, and networks, case studies have the force of example. The often rich and detailed narrative of case studies makes visible the complexities, nuances, and ambiguities of real life, without simplifying such realities (Law 2004; Flyvbjerg 2006).

After a first exploration, we selected two research sites, both located in Bhaktapur District east of Kathmandu (Figure 4.1), for field research to capture various types of hydro-social change related to urbanization, located in various waterscapes and involving various technologies. The first area, Golmadi, is in the Bhaktapur Municipality. On the edge of the old town

[1] The second author (Roth) was involved as research supervisor, while the other authors (Dongol, Shrestha, and Yakami) acted as local resource persons, giving feedback and sometimes joining in field visits and interviews.

FIGURE 4.1
Map of study locations.

center and enclosed by buildings, the agricultural fields in this area are quite small. Farmers irrigate with water coming directly from a sewage pipe crossing the fields. The extreme condition of this location where raw, completely untreated wastewater serves as irrigation source makes it a very interesting study area. The second area, Dadhikot,[2] is in the Suryabinayak Municipality, established in 2017. It is a bigger area situated along the Hanumante River, rapidly urbanizing but retaining many rural characteristics, especially its land use for (irrigated) agriculture. Farmers rely on more sources here. An irrigation canal serves the farmers closer to its source, a *kulo* (surface irrigation system). Groundwater wells are an alternative or additional source, but do not always provide water due to the declining groundwater table. Finally, farmers pump heavily polluted water from the Hanumante River when the other sources fail (Table 4.1).

Main fieldwork methods were field observations, participant observation, and interviews with farmers, government officials, administrators, and other relevant officials. We held short interviews at both sites with almost all farmers who owned or worked land to get an overall idea of the farmer population and their characteristics. We held in-depth interviews with farmers

[2] Before municipal declaration Dadhikot was part of Dadhikot Village Development Committee (VDC), a separate administrative unit.

TABLE 4.1

Overview of Basic Characteristics of the Selected Study Locations

Indicators	Golmadi (Bhaktapur)	Dadhikot (Suryabinayak)
Household income	Mostly diversified	More than half of households rely on diversified incomes
	Only two households depend completely on agriculture	Almost half of households depend only on agriculture
Plot size	~700 m^2 on average (31–2,000 m^2)	~900 m^2 on average (55–4,000 m^2)
Rent costs (USD/ropani (~500 m^2)/season)	115–140	32
Land tenure	25% households renting land (7)	44% households renting land 35% renting and owning land
Water sources used for irrigation	Sewage pipe	Hanumante River Irrigation canal Wells
Type of wastewater	Concentrated	Diluted
Investments in agriculture	Low (not paying for water, no wells, no pumps)	High (construction of wells, use of pumps)
Agricultural Department	No involvement	Registered groups of farmers get Agricultural Department support

selected for their positions relative to water sources, specific views, experiences with and knowledge of relevant issues, and position in the community, or through snowball sampling. We recorded the interviews in field notes or by audio recordings; photographic material helped in visualizing research choices and experiences. We conducted interviews in Nepali or Newari with the help of translators.[3]

4.3 Development of Wastewater Irrigation in Kathmandu Valley

Kathmandu Valley is among the most productive agricultural areas in Nepal, with rice, wheat, and maize yields significantly higher than national averages (HMG [His Majesty's Government of Nepal] and United States Agency for International Development [USAID] 1986). The ancient canal irrigation systems known as *rajkulo* (royal canal), fed by the Bagmati River or its tributaries,

[3] Thanks are due to the following students of Nepal Engineering College: Sarita Shrestha, Nishant Shrestha, Jyoti Dahal, and Saroj Malakar.

played an important role in enriching this agricultural potential.[4] Since the 1960s the agricultural landscape of the valley started urbanizing.[5] Unplanned urban expansion and influx of people from the 1980s led to a growing production of wastewater. The area introduced four large-scale and centralized wastewater treatment systems in the 1970s and 1980s, and another in the early 2000s.[6] These systems, targeted for treatment of urban sewerage, were unable to deal effectively with the waste of expanding residential, commercial, and industrial establishments, resulting in disposal of untreated waste into the Bagmati River and its tributaries (Shukla et al. 2012; KUKL [Kathmandu Upatyaka Khanepani Limited] 2013; Jha and Bajracharya 2014).

Over the decades, the area constructed few public and private decentralized wastewater treatment systems (DEWATs). Municipal efforts have largely failed in managing the increasing volume of domestic and industrial effluents. For example, the integrated waste management system in Sano Khokana, constructed to use treated effluent in nearby agriculture, fails to treat wastewater (Shrestha 2016). Although the practice of recovering nutrients in wastewater exists in traditional agricultural systems (Shrestha 2011), increasing pollution of traditional irrigation sources left farmers with no option but to use polluted water. Farmers practicing wastewater irrigation get it directly from municipal sewage, from rivers with diluted wastewater, or from wastewater ponds and pools developed in urban, peri-urban, and rural areas (Shukla et al. 2012). Farmers use wastewater especially for cash crops which require frequent and reliable irrigation in this increasingly water-deficit area.[7] A study conducted in Harisiddhi revealed that farmers accept wastewater because of its easy access and year-round availability (Baidya 2016): it has created opportunities for cultivation of high-value crops, even in the dry season, to fulfil the growing need for vegetables in urban areas.

However, high nutrient levels in wastewater can also cause skin and other health problems, and crop damage: effluents in the valley register high turbidity, microbial, and nutrient levels (Baidya 2016), all critical parameters for irrigation water. Although wastewater use is largely informal and unrecognized by the government, it is rapidly spreading (Rutkowski et al. 2007; Sada 2010; Shukla et al. 2012). The area is developing new treatment plants,[8] but untreated wastewater irrigation practices are likely to increase

4 The population widely uses the Bagmati River and its tributaries for drinking, irrigation, industrial, and other purposes in the valley (GoN/NTNC 2009).

5 The urban population in the valley increased six-fold in the last 60 years, from 218,092 people in 1961 to 1,426,641 by 2011. The built-up area increased by 134 percent, from 24.54 square kilometers in 1989 to 57.32 square kilometers in 2006 (Shrestha et al. 2014).

6 The total theoretical capacity of these plants is 36.3 megaliters per day (Rutkowski et al. 2007).

7 Water demand has reached 375 megaliter per day while supply is 118 megaliters per day (wet season) and 73 megaliters per day (dry season) (KUKL 2016).

8 Kathmandu Valley Wastewater Management Project supported by ADB underway since 2013/14.

in the rapidly urbanizing valley (World Health Organization 2006). This situation makes case studies on its emergence and experiences of wastewater users timely and relevant. While most research deals with wastewater from an instrumental perspective, focusing on its role in food and environmental security and poverty alleviation (Raschid-Sally and Jayakody 2009), we need a deeper understanding of the daily practices around wastewater irrigation as a solid ground for policy-makers in their future management of peri-urban waterscapes.

4.4 Two Cases of Wastewater Irrigation

4.4.1 Bhaktapur

4.4.1.1 History of a Sewage Pipe

In Bhaktapur, the sewage pipe there used to be a *rajkulo* that provided potable water for the population of Bhaktapur. The canal was the main source of water for the whole town and stone spouts and ponds distributed the water further. When flowing through farmland, farmers also used the water as a source of irrigation water (Gautam 2015).

Between 1974 and 1986 the Bhaktapur Development Plan (BDP), locally known as the German Project, promoted urbanization and industrial development in the city. The project, arranged between the Government of the Federal Republic of Germany and His Majesty's Government of Nepal, with the technical support of the (then) German Association for Technical Cooperation (GTZ; Deutsche Gesellschaft für Technische Zusammenarbeit), focused on improving hygienic conditions, boosting public services and economic activities in the city, restoring buildings, and creating institutions to maintain the improvements achieved (Matthaeus 1988). Construction of the sewage network, sewage treatment plants, public toilets, and promotion of installation of private toilets in each household were part of this project (Matthaeus 1988).

After this intervention, sewage water from Bhaktapur started flowing through the *rajkulo*. Now a brick pipe with a diameter of about 3 m, also used as a pedestrian passage, crosses the fields in Golmadi. Eventually it drains into the Hanumante River. The pipe has one manhole to reach the interior of the pipe and other small openings all along it on both sides and close to the ground. Farmers use the water from this pipe to irrigate their fields. The original plan included the construction of two treatment plants in two locations, Sallaghari and Hanumanghat. However, today these are either malfunctioning or not in use anymore (Sada 2010). The centralization of management of the wastewater flowing out of the city caused a higher level of pollution in the canals flowing into the river and the *rajkulos*, disrupting the pre-existing household-level water management systems in the area and deeply affecting the livelihoods of farmers (Sada 2010) because they now depend on it. A farmer said: "*we have never been told not to use the wastewater from the municipality. But even if people were told so, they would not agree*" (Golmadi, January 1, 2016). Although Nepal considers this practice illicit, the municipality, aware of the lack of alternatives for farmers in the area, does not intervene against it.

For government and municipality officials, introduction of the sewage system was a sign of development and its malfunctioning caused by poor maintenance and farmers' irrigation practices from the pipe. According to the farmers it meant an improvement of living conditions, also allowing them to have greater water availability throughout the year. Since its construction, farming in Golmadi has undergone an evolution: the water flow has increased but water quality and access have changed. Farmers have introduced new cropping patterns because year-round water availability allows them to grow vegetables also in the dry season. "*Before the German project in this area we used to cultivate only wheat and paddy* [in the monsoon season], *there was not enough water available* [*for other crops*]" (farmer, Golmadi, January 19, 2016). "*If the pipe hadn't been constructed, this land would not be suitable for cultivation, due to lack of water.* [...] *If the sewage had not been built, we would have to pump the water from the river, which is expensive.* [...] *We can now irrigate only thanks to the wastewater*" (farmer, Golmadi, January 13, 2016). Thus, thanks to wastewater, farmers can produce crops all year close to markets, their homes, and customers. The new water flows are influencing cropping patterns, market supplies, and income sources, showing once more the interconnectedness between water flows and social and economic changes (Table 4.2).

Most farmers state that during the drier summer months the wastewater is bad both for the soil and crops which become yellow and often die. However, most of them consider the health hazards and poor water quality minor issues that they need to overcome anyway for lack of alternatives: "*for an alternative to wastewater, we have to buy a field somewhere else or dig a well, which are both expensive options.* [...] *Everyone would like to dig a well but they cannot afford it*" (farmer, Golmadi, December 29, 2015).

The sewage pipe has also affected access to water for the farmers. With the *rajkulo* they only had to block the open canal to divert the water flow. A young farmer performing this labour: "*I do the blocking of the pipe with three bags of sand. There's no light so I need a torch. It's dangerous, the soil is slippery. It only takes 5 min to do it.* [...] *The last one to irrigate will unblock the main sewage pipe*" (Golmadi, January 22, 2016). Men usually do this, as a woman explains: "*I don't like to do it, it is smelly and dirty*" (Golmadi, January 20, 2016). Then, the water flows to the fields through open canals diverted by farmers often walking barefoot or digging the soil with bare hands: some farmers got skin rashes because of the contaminated water.

On the eastern side of the sewage canal the water only serves fields adjacent to the pipe, so each farmer has developed his or her own system, for

TABLE 4.2

Characterization of Farmers in Golmadi, Bhaktapur

Characterization of Farmers in Golmadi, Bhaktapur		
# interviewed farmers	39	
Household size	2–11 members	
Income	Mostly diversified (manufacturing, sales and retail occupations, transport and construction, private businesses, a few professionals and people working abroad)	
Who takes care of the fields	Mostly middle-aged people (parents/grandparents)	
	Mostly female	
Paid farm labour	Only two households. Mostly labour exchange practice	
Home distance	Average 5–10 min' walk	
Land size	~700 m² on average (31–2,000 m²)	
Land ownership	Mostly inherited	
Land tenure	Only six households. Payment in money or 50% output	
Rent prices	USD 1.2–2.2/1 aana (31.8 m²)	
Production	**Winter**	**Summer**
	Vegetables	Rice
	Potatoes	Wheat
		Chili
		Eggplant
Production purpose	Mostly self-consumption, market	

Year-round agriculture benefit from wastewater usage

example, by connecting a small pipe to the openings in the sewage pipe. When not in use, the farmer closes the pipe with a brick. When the farmer moves the brick the water immediately flows out. Some farmers use buckets to irrigate their plots, taking the water from the open canals that carry the wastewater through the fields. One farmer, whose house is in the fields, irrigates her small plots with water from a newly installed tap. An elderly farmer fetches water at wells and ponds near the house when the wastewater is not enough. Two farmers also pump water from the Hanumante River: *"most people don't use a pump but wastewater, even if its quality is not good. It is difficult for one person alone to bring the pump and apply the water, and on these fields, there is mostly only one person working [per plot]. Additionally, it is expensive, and people don't own pumps here"* (farmer, Golmadi, January 22, 2016).

There are no strict rules that set a specific irrigation order among farmers. Unwritten and even unspoken social agreements to a large extent guide wastewater use practices: *"when you need the water you can just take it, there is no form of organization around it. Usually there is enough for everyone, but you have to wait for your turn, until your neighbour is done with it"* (farmer, Golmadi,

December 28, 2017). Some farmers agree in stating that upstream farmers are the first ones to irrigate, and this sometimes causes damage to their crops downstream because they must sow or irrigate their crops too late. A combination of factors, however, makes this imperfect system suitable for farmers in Golmadi. The fact that the area is not too large probably reduces the impact of the upstream-downstream relations. In addition, this method is the only way of accessing irrigation water for many farmers in the area.

Although farmers tend to go for individual water access solutions that best fit their financial and organizational capacities, awareness of each other's interests and forms of cooperation do play a role. Cooperation of farmers in maintenance and cleaning is necessary for retaining access to the sewage water because with time canals and pipes get clogged if not properly maintained. The maintenance takes place twice a year, during the dry season or at the beginning of the rainy season. The system is practical and flexible, tailored to the farmers' needs and the individual requirements of various people. Landownership and seniority are the leading forces behind the system: the most vocal farmers are those who have owned the larger land plots for a longer time. The system does not require the presence of farmers owning smaller plots because they can pay some money depending on their willingness, the amount of land owned or rented, and duration of the maintenance work. The few farmers not contributing in any way are usually not benefiting from the canal system. The existence of a form of social organization around the wastewater irrigation system and the sewage pipe allows many farmers to benefit from this source.

4.4.1.2 Golmadi: Wastewater as an Opportunity

In this dynamic combination of rural and increasingly urban characteristics found in Golmadi, the advancing city is literally enclosing the farmland. Being cut off from their traditional surface irrigation canals increasingly constrains the farmers. Within these constraints, however, construction of the sewage system has also brought new benefits, primarily year-round access to irrigation water. Although water quality is low, it brought new livelihood options. Thus, the farmers keep working their land aided by their connection to wastewater. Awareness of the changes around them does not make farmers less willing to continue farming because it is a necessity for most of them. Even while the city slowly encloses them, they do not seem prone to drop their shovels.

4.4.2 Dadhikot

In Dadhikot a *"drastic land use change in the past 5 or 6 years"* (farmer, Dadhikot, January 26, 2016) has taken place. Buildings, shops, brick kilns, and other enterprises and infrastructure are arising everywhere, and land prices are increasing exponentially. Farmers are seeing land and water encroached by

TABLE 4.3

Characterization of Farmers in Dadhikot

Characterization of Farmers in Dadhikot, Suryabinayak		
# interviewed farmers	51	
Household size	3–16 members	
Income	More than half diversified (transport and construction, work abroad, professionals, manufacturing, office work). Others agriculture only.	
Who takes care of the fields	Mostly husband and wife	
Paid farm labour	Almost half of households have seasonal/yearly employees	
Home distance	On field/bordering municipality	
Land size	~910 m^2 average (55–4,030 m^2)	
Land ownership	More than half farmers own land. Mostly bought.	
Land tenure	Almost half. Payment in money or 50% output	
Rent prices	32 euros/ropani (500 m^2)/6 months	
Production	**Winter**	**Summer**
	Vegetables	Rice
	Potatoes	Wheat
		Chili
		Eggplant
Production purpose	Mostly market	

A declining irrigation source: the *kulo*

new inhabitants. The first and most contested irrigation source in Dadhikot is the *kulo*, the irrigation canal. It comes from a seasonal stream, the Chakhu Khola, that farmers for many years diverted into their fields with a dyke of sandbags. The kulo water availability shows seasonal variations in accordance with the hydrological seasons of the Bagmati River Basin: a dry premonsoon season from March to May, the rainy monsoon season from June to September, and the post-monsoon season with little rain from October to February (Sharma and Shakya 2006) (Table 4.3).

Aside from changes in rainfall patterns and reduced source water availability, farmers specifically blame population increase for the reduction of water availability in the *kulo*: *"There are houses everywhere and everyone draws water from the kulo. There used to be enough water every year in Falgun [February–March] but this year there is no water and there has been no rainfall. It has decreased compared to last year"* (farmer, Dadhikot March 14, 2016). This change is intensifying water distribution inequalities. Farmers closer to the source mostly get enough water; however, those further away get little or no water. Several farmers have lost access to the *kulo* because of buildings among the fields. One farmer says: *"I don't get water anymore … because there are houses everywhere, both of people with legal permission to build a house and of people who… have taken the land around the kulo illegally*" (farmer, Dadhikot, February 11, 2016).

After the rainy season, farmers conduct the maintenance around the *kulo*. Around 20 to 30 farmers, mostly men, gather with some of them leading the process. Whoever cannot join pays money to a treasurer, who manages the expenses. One farmer states: "*often the landowners only grow paddy [in the summer season], so the tenant is responsible for maintenance*" (Dadhikot, January 6, 2016). The farmers have decreased the frequency of maintenance over the years. One farmer stated: "*50 or 60 people were participating previously, nowadays only 30. Less people are involved in agriculture, and there are more houses around*" (Dadhikot, February 21, 2016). Although the maintenance system is still active, it is becoming more informal. Reduced access to water for farmers and the decrease of people involved in agriculture are roots of these changes.

The first-come first-served principle seems to be leading in using kulo water, together with mutual understanding, daily agreements between farmers, and respect for each other's turns. Sometimes conflicts arise: "*There are conflicts among farmers but not serious ones, and always solved by talking. Usually upstream - downstream issues* […]. *Upstream people use more water and block the kulo when irrigating, so that people downstream don't get enough*" (farmer, Dadhikot, February 23, 2016). Wide changes in water availability fuel such issues. Downstream users are often looking for alternative water sources or practices, such as irrigating at night or drawing water directly from the kulo source, the Chakhu Khola. Some are hoping for governmental support for repairing the *kulo*.

4.4.2.1 New Irrigation Practices, Technologies, and Problems

The increasing water stress is forcing many farmers to turn to other water sources, such as wells. A few scattered ones are in the fields, built by private owners or with funding of the Agricultural Department. Farmers value well water for its quality and use it mostly for washing vegetables and drinking. Only a few farmers use groundwater for irrigation. One farmer says: "*10 or 12 years ago I used well water to irrigate, but not anymore. It used to take 4 h to irrigate with well water. Now it takes 7 h and too much fuel*" (Dadhikot, February 7, 2016). Although farmers share most wells, groundwater is undergoing a privatization process. Some farmers are not allowing neighbors to benefit from their wells anymore, while new households are constructing wells for their own use. The growing number of wells and increasing water consumption are causing a lowering of the groundwater table. Often in the dry season many wells run dry. The reliance of many new houses on well water deprives farmers of this clean water source which is switching from agricultural to domestic use.

Most farmers are turning to the Hanumante River as main irrigation water source. While farmers have always used the river for this purpose, the

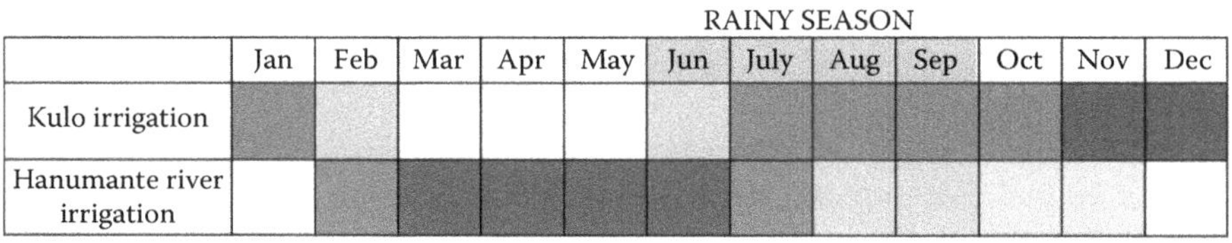

FIGURE 4.2
Irrigation sources calendar.

number of users is increasing. Water quality has changed drastically since the creation of centralized sewage systems that dispose increasing quantities of treated water into the river: *"We used to drink the river water, but now we cannot even touch it"* (farmer, Dadhikot, February 5, 2016). Many farmers incur health problems, mainly skin rashes and itching, from contact with river water while irrigating. Hence, whoever has access to the *kulo* prefers that source: *"in April the smell of the river water is terrible. We only use it when we have no other option. If we use too much diluted wastewater the vegetables become yellow"* (farmer, Dadhikot, January 6, 2016). Figure 4.2 shows the use of different water sources by farmers: darker colors mean higher water availability and more farmers relying on that water source, and so also more competition.

The use of river water and groundwater has introduced new technology: water pumps. Half of the interviewed farmers own a pump and the others rent one. Often, pump-owners help their customers placing the pipes and operating the pump. One farmer states: *"there is no organization for getting the water from there* [*the river*]*: whoever can afford a pump, pumps the water. Others pay the pump owner 250 rupees per hour"* (Dadhikot, December 23, 2015). Thus, with new technology new differences are arising. These differences gain relevance especially in the dry season, when farmers need to irrigate more often. Another farmer said: *"I can't irrigate whenever I want to, as I depend on availability of the pump.* [...] *Some people irrigate at night, with nobody around they can use as much water as they want"* (Dadhikot, February 21, 2016).

4.4.2.2 Dadhikot: Growing Pressures on Irrigated Agriculture

Decreasing government support for farmers, an increasing number of young people seeking a future outside agriculture, land use changes, and the weakening of water rights and maintenance mechanisms in the irrigation systems are all signs of urbanization in Dadhikot. Houses and households are increasing, rivers are turning into sewers, and irrigation canals are disappearing from the map. In this rapidly urbanizing landscape, increasingly also influenced by a changing climate, farmers try to uphold their farming-based livelihoods.

4.5 Analysis: Multiple Agencies, Assemblages, and Changing Access to Water

This chapter has taken the changing water flows and water infrastructures of the two case study areas as its point of departure, approaching the irrigated areas as assemblages of a diversity of human and non-human elements, including farmers and farmer organizations, changing land uses and agricultural practices, water flows, spaces, rivers, canals and water infrastructure, technologies, policies and flows of funding, land and water rights, and access to water. This change highlights the peri-urban dynamics in various ways: the provisional and contingent character of the hydro-social (trans-)formations involved, the multiple human and non-human agencies involved in the processes of shaping and reshaping them from heterogeneous elements, and their emergent yet also coherent character (Anderson and McFarlane 2011). It also highlights that these processes of assembling, reassembling, or both shaped by, and re-shaping the ways in which farmers define water rights and hydraulic property, and create access to water (Ribot and Peluso 2003; Roth et al. 2015), which we will clarify in the following analysis.

We can consider the sewage pipe in Bhaktapur as an interface (Long 2001) or as a key material element in an assemblage-involving multiple, human, and non-human agencies. In a chain of agencies, farmers are the actors who reshaped the technology to use it for a purpose not intended. Using the manhole, needed to control the functioning of the pipe, the farmers restored the irrigation system they used when the sewage pipe was an open canal. The clash between multiple values and realities comes into being exactly at the location of the sewage pipe and the area around it. First, conflicting views are emerging concerning the characterization of the peri-urban with farmers on one side and any of the government agencies or municipality officers on the other side. For most farmers, this area is the place where they have always farmed. Their need for water to irrigate the crops they depend on for their livelihoods drove them to use the new water source, despite its new function and, hence, lower water quality. The farmers' motives for holding onto their farmland are stronger than the many changes taking place: livelihoods, family ties, and lack of alternative options. Their priority is not cleanliness of the water, but water availability. It even became an opportunity: a constant water flow even in the dry season. For farmers in Golmadi, these reasons are strong enough to make them stay and continue their agricultural practices using the sewage water and turning this area one of the last agricultural strongholds in the city.

The new water source creates new definitions of water rights and legitimate access. This legal aspect has pushed farmers to participate in the social organization around the sewage-based irrigation system. Strong organizational ties between farmers have developed around the sewage pipe. Labour for maintenance, time, and money invested, materials used, and sticking to the locally developed agreements are necessary conditions for everyone to retain recognized access to wastewater; thus, the system shows characteristics of hydraulic property (Coward 1986, in Roth et al. 2015). Even though the system is based on soft rules, almost everyone participates personally or with monetary support because this way they can benefit from the resource in a way regarded as legitimate locally (Ribot and Peluso 2003; von Benda-Beckmann et al. 2006).

The Dadhikot case involves changing water sources, new water sources, and new actors in the network. The *kulo* is still the basis of farmers' customary water rights. Even though increasingly flexibly organized, the right to use irrigation water continues to be tied to material and labour investments in construction and maintenance, a form of hydraulic property. However, this arrangement is no longer a guarantee of actual water access (Ribot and Peluso 2003; Roth et al. 2015). While less people now are investing in *kulo* maintenance, house owners are increasingly using the system. The latter, however, do not participate in the process that would allow them, according to customary rules, to benefit from the canal. They do not feel tied to the farmers' customary norms and use their ability to directly access the source.

New landowners also claim property rights to groundwater according to the 1992 Water Resources Act gained along with land owned. For them, the well is their own property, not a communal resource. These new water users are carrying with them a whole new set of norms related to the association between land control and the right to use groundwater. As this new rights-based mechanism arises—farmers claiming rights to groundwater because they own the land—some of the structural-relational mechanisms (Ribot and Peluso 2003) through which farmers based their access to water are also affected: the water source is declining or depleted and so farmers need deeper wells and new and increasingly powerful pumping technology to access the groundwater. This situation creates new divisions between those who retain access to groundwater and those who gradually lose their ability to access it.

The increasing number of farmers shifting to river water is changing the nature of the assemblage from one where the whole area served farming purposes and farmers strongly tied to the *kulo*, its organizational arrangements, and system-based water rights, to an assemblage where the kulo is becoming part of a more heterogeneous peri-urban waterscape. In the latter farmers are creating new networks around alternative water sources, using new technologies, infrastructure, and resources (pumps, pipes, money, and fuel), different types of water rights, more flexible and ad-hoc forms of access to water, and a less structured social organization around it. The process creates new discrepancies between farmers. The water flows that bring these changes into being are not socially neutral, as is also stressed in the urban political ecology of water literature (Swyngedouw et al. 2002). Another focal point in political ecology literature is the scalar dimension of these flows between urban and peri-urban areas. The case studies show how urbanization is encroaching on relatively clean peri-urban water sources and creating new flows of polluted water for urban or peri-urban re-uses, such as in peri-urban agriculture in the cases we discussed. The actors affected by this quality shift are those at the margins of urbanization, farmers who do not have other options but to keep on using whatever is available around them.

4.6 Conclusion

Building on insights from the political ecology of water, actor-network theory, assemblage theory, and anthropology of law in this chapter, we have tried to understand farmers' changing irrigation practices that are increasingly based on wastewater or heavily polluted water in the dynamic peri-urban environment of Kathmandu Valley. We brought forward the perceptions, motives, and choices (or lack of them) behind these practices as they are developing in two irrigation water use settings. The two cases of changing irrigation practices—Golmadi, Bhaktapur case (wastewater irrigation from a sewage system) and Dadhikot case (pumping of polluted water from Hanumante river)—mainly differ in the combinations of water sources farmers can still rely on as urbanization is advancing. In both cases, urbanization is attracting new infrastructures and people, introducing new notions of water rights and land rights, and new organizing practices to create, maintain, and control access to water to enhance water security. Fundamental discrepancies in the perception of these changes are emerging, further enhancing the growing peri-urban water stress for some while creating opportunities for others.

The analysis highlights the importance of infrastructure in these shifting peri-urban interfaces, and the central role played by the agency of different human and non-human actors in shaping and reshaping the reality around them. Confronted with growing land and water stress in an urbanizing landscape, farmers re-interpret infrastructure, introduce new technologies, change cultivation and irrigation practices, and find new ways of establishing, maintaining, and strengthening their water rights, water access, and water security. Irrigation practices are constantly changing along with urbanization, water rights are becoming less explicit, and access to water more ad-hoc. The case studies show how unpredictable and context-specific farmers' responses to changes in land and water are. To understand how to deal with these changes, it is essential to start from an analysis of daily farming practices in peri-urban fringes. By looking at changing flows and assemblages, we focus not only on the specific geographical study area, but also on the concept of peri-urban as a process, a constantly evolving network. We followed the changing water flows and relationships, highlighting the relevance of looking at the peri-urban as a waterscape (or wastewaterscape). Such research might give new insights in the changes taking place and expand our knowledge and understanding of peri-urban processes.

References

Allen, J. 2011. Powerful assemblages? *Area* 43(2):154–157.

Anderson, B., and C. McFarlane. 2011. Assemblage and geography. *Area* 43(2): 124–127.

Bartlett, R., L. Bharati, D. Pant, H. Hosterman, and P. G. McCornick. 2010. *Climate Change Impacts and Adaptation in Nepal* (Vol. 139). Colombo, Sri Lanka: IWMI.

Baidya, M. 2016. *Potential of Wastewater Use for Water Security in Irrigated Agriculture: Case of Harisiddhi Wastewater Treatment Plant, Nepal.* Master's thesis. Bhaktapur, Nepal: Pokhara University, Nepal Engineering College.

Budds, J. 2009. Contested H_2O: Science, policy and politics in water resources management in Chile. *Geoforum* 40:418–430.

de Leeuw, J. 2014. *Multiple Perspectives on the Use of Wastewater in Agriculture. A Study among Farmers and Customers on the Use of Wastewater in Irrigated Vege Cultivation in the Kathmandu Valley, Nepal.* Master's of Science thesis. Wageningen, the Netherlands: Wageningen University.

Dhakal, S. 2011. *Land Tenure and Agrarian Reforms in Nepal.* Kathmandu, Nepal: Community Self-Reliance Center.

Flyvbjerg, B. 2006. Five misunderstandings about case-study research. *Qualitative Inquiry* 12(2):219–245.

Gautam, D. P. J. 2015. Distributional and cascades reusability aspects of indigenous water management system of Bhaktapur City, Nepal. *Paper presented at the 14th International Symposium on New Technologies for Urban Safety of Mega Cities in Asia,* Vol. 14, Kathmandu, Nepal.

GoN/NTNC. 2009. *Bagmati Action Plan (2009–2014).* Kathmandu, Nepal: Government of Nepal and National Trust for Nature Conservation.

HMG and USAID. 1986. *Kathmandu Valley Urban Land Policy Study.* Kathmandu, Nepal: His Majesty's Government of Nepal and The United States Agency for International Development.

Iaquinta, D. L., and A. W. Drechser. 2000. Defining peri-urban: Understanding rural–urban linkages and their connection to institutional contexts. *Paper Presented at the Tenth World Congress of the International Rural Sociology Association,* Rio de Janeiro, Brazil.

Jha, A. K., and T. R. Bajracharya. 2014. Wastewater treatment technologies in Nepal. *Proceedings of IOE Graduate Conference,* Kathmandu, Nepal.

Karpouzoglou, T., and A. Zimmer. 2016. Ways of knowing the wastewaterscape: Urban political ecology and the politics of wastewater in Delhi, India. *Habitat International* 54:150–160.

KUKL. 2013. *NEP: Kathmandu Valley Wastewater Management Project, Draft Initial Environmental Examination.* Kathmandu Upatyaka Khanepani Limited, Ministry of Urban Development, Government of Nepal for the Asian Development Bank.

KUKL. 2016. *Annual Report Eighth Anniversary.* Kathmandu, Nepal: Kathmandu Upatyaka Khanepani Limited.

Latour, B. 1999. On recalling ant. *The Sociological Review* 47(S1):15–25.

Latour, B. 2005. *Reassembling the Social: An Introduction to Actor-Network-Theory.* Oxford, UK: Oxford University Press.

Law, J. 1992. Notes on the theory of the actor-network: Ordering, strategy, and heterogeneity. *Systems Practice* 5(4):379–393.

Law, J. 2004. *After Method: Mess in Social Science Research*. London, UK: Routledge.

Li, T. M. 2007. Practices of assemblage and community forest management. *Economy and Society* 36(2):263–293.

Long, N. 2001. *Development Sociology: Actor Perspectives*. London, UK: Routledge.

Matthaeus, H. 1988. *Bhaktapur, Nepal. Integrated Urban Renewal and Development. A Review of 12 Years Development Cooperation between His Majesty's Government of Nepal and the Federal Republic of Germany*. Eschborn, Germany: Deutsche Gesellscahft fuer technische Zusammenarbeit.

Narain, V. 2009. Growing city, shrinking hinterland: Land acquisition, transition and conflict in peri-urban Gurgaon, India. *Environment and Urbanization* 21(2):501–512.

Narain, V. 2014. Whose land? Whose water? Water rights, equity and justice in a peri-urban context. *Local Environment* 19(9):974–989.

Narain, V. 2016. Introduction. In *Water Security in Peri-Urban South Asia. Adapting to Climate Change and Urbanization*, (Eds.) V. Narain and A. Prakash, pp. 1–32. New Delhi, India: Oxford University Press.

Narain, V., and A. Prakash, Eds. 2016. *Water Security in Peri-Urban South Asia. Adapting to Climate Change and Urbanization*. New Delhi, India: Oxford University Press.

Narain, V., M. S. A. Khan, R. Sada, S. Singh, and A. Prakash. 2013. Urbanization, peri-urban water (in) security and human well-being: A perspective from four South Asian cities. *Water International* 38(7):930–940.

Raschid-Sally, L., and P. Jayakody, Eds. 2009. *Drivers and Characteristics of Wastewater Agriculture in Developing Countries: Results from a Global Assessment* (Vol. 127). Colombo, Sri Lanka: IWMI.

Raschid-Sally, L., I. M. Kengne, V. A. Nguyen, and D. Endamana. 2008. Wastewater use in high rainfall Riverine cities: Comparisons from Cameroon, Nepal and Vietnam. In *Water Reuse: An International Survey of Current Practice, Issues and Needs*, Vol. 20, (Eds.) B. E. J. Cisneros, B. Jiménez, and T. Asano, pp. 544–557. London, UK: IWA Publishing.

Ribot, J. C., and N. L. Peluso. 2003. A theory of access. *Rural Sociology* 68(2):153–181.

Roth, D., R. Boelens, and M. Zwarteveen. 2015. Property, legal pluralism, and water rights: The critical analysis of water governance and the politics of recognizing "local" rights. *Journal of Legal Pluralism and Unofficial Law* 47(3):456–475.

Rutkowski, T., L. Raschid-Sally, and S. Buechler. 2007. Wastewater irrigation in the developing world—Two case studies from the Kathmandu Valley in Nepal. *Agricultural Water Management* 88(1):83–91.

Sada, R. 2010. *Processes and Consequences of Degradation of Hanumante River: Religious, Cultural and Livelihood Impacts*. Master's thesis. Bhaktapur, Nepal: Pokhara University, Nepal Engineering College.

Sharma, R. H., and N. M. Shakya. 2006. Hydrological changes and its impact on water resources of Bagmati Watershed, Nepal. *Journal of Hydrology* 327(3):315–322.

Shrestha, J. 2011. *Traditional Practices and Knowledge System in Integrated Wastewater Management in Kathmandu Valley: Case Study of Khokana VDC*. Master's thesis. Bhaktapur, Nepal: Pokhara University, Nepal Engineering College.

Shrestha, D. 2016. *Treatment Efficiency and Socio-Environmental Acceptance of Constructed Wetlands in Kathmandu Valley*. Master's thesis. Bhaktapur, Nepal: Pokhara University, Nepal Engineering College.

Shrestha, A., and R. Sada. 2013. Evaluating the changes in climate and its implications on peri-urban agriculture. *Merit Research Journal of Agricultural Science and Soil Sciences* 1(4):48–57.

Shrestha, A., R. Sada, and L. Melsen. 2014. Adapting to peri-urban water insecurity induced by urbanization and climate change. *Hydro Nepal: Journal of Water, Energy and Environment* 14:43–48.

Shukla, A., M. Prajapati, R. Sada, and A. Shrestha. 2011. *Water Security in Peri-Urban South Asia: Adapting to Climate Change and Urbanization*. Scoping Study Report. Hyderabad, India: SaciWATERs.

Shukla, A., U. R. Timilsina, and B. C. Jha. 2012. Wastewater production, treatment and use in Nepal. http://www.ais.unwater.org/ais/pluginfile.php/232/mod_page/content/134/Nepal_CountryPaper.pdf (accessed May 16, 2017).

Swyngedouw, E. 1999. Modernity and hybridity: Nature, regeneracionismo, and the production of the Spanish waterscape, 1890–1930. *Annals of the Association of American Geographers* 89:443–465.

Swyngedouw, E., M. Kaika, and E. Castro. 2002. Urban water: A political ecology perspective. *Built Environment* 28(2):124–137.

von Benda-Beckmann, F., K. von Benda-Beckmann, and M. G. Wiber. 2006. The Properties of Property. In *Changing Properties of Property* (Eds.) F. von Benda-Beckmann, K. von Benda-Beckmann, and M. G. Wiber, pp. 1–39. New York: Berghahn Books.

Water and Energy Commission Secretariat. 2011. *Water Resources in Nepal in the Context of Climate Change*. Kathmandu, Nepal: WECS.

5

Climate Migration and Flood Related Disasters

Anna Wesselink, Jeroen Warner, Michelle Kooy, Parvin Sultana, Paul Thompson, and Abu Syed

CONTENTS

5.1 Introduction 71
5.2 Floods and People in Bangladesh 72
5.3 The Rise of Climate Change Migration as a Global Issue 74
5.4 Climate Change and Migration in Bangladesh 78
5.5 An Outlook: Prudent Governance of Migration 82
References 83

5.1 Introduction

Studies often present Bangladesh one of the countries where the impact of climate change is imminent (or already started) and hard-hitting. The main reason for this assessment is Bangladesh's location in the delta formed by the Brahmaputra, Meghna, and Ganges rivers. More specifically, studies expect climate change will increase the frequency and magnitude of flooding through increased river runoff and sea level rise (Auerbach et al. 2015). This trend is also expected to happen in other deltas in the world, but with Bangladesh's high population density and high levels of poverty, the consequences to human life and society are often portrayed as cataclysmic and potentially leading to unmanageably large numbers of so-called climate refugees, "the human face of global warming" (Salauddin and Ashikuzzaman 2011; Ahsan et al. 2014). Regular reports of recurring destructive flood events in Bangladesh and associated human misery reinforce the notion that the country will not be able to cope with climate change since such events would only increase in size and frequency.

Many of the current governance debates on how to respond to climate change in Bangladesh, and elsewhere, assume that climate change-induced migration will happen (or is already happening) and that climate change is the main driver for (increased) migration (Alam 2004; Afifi and Warner 2008; Biermann and Boas 2008). However, detailed studies reveal that drivers for migration are more complex (Joarder and Miller 2013; Penning-Rowsell et al. 2013; Yasmin and Ahmed 2013; Etzold et al. 2014; Islam et al. 2014). These detailed studies find that push factors leading to migration may include environmental changes among many other reasons, but pull factors such as employment opportunities at the destination also play a major role. Despite this growing body of evidence, the assertion that Bangladesh is going to face a major crisis due to climate change (as if the same problems do not already exist without climate change) is widespread and influential, and we expect it will continue to be so. We assert that this portrayal is not primarily based on established and emerging insights but stems from key interests in the so-called development industry (Hancock 1989; Powell and Seddon 1997) and associated research and governmental arenas. This means that multi-billion dollar investments in projects and programs are based on speculative and simplistic future projections rather than on evidence about actual lived experiences.

We will first sketch the historical dynamics of floods and people in Bangladesh. We then outline the development of global narratives on the relation between climate change and migration before discussing how projects and programs view and act upon this relationship in Bangladesh. We conclude with a few pointers towards more realistic and prudent governance of migration.

5.2 Floods and People in Bangladesh

The Ganges-Brahmaputra-Meghna delta is the largest delta in the world. It encompasses at least 280 river channels and covers most of the country of Bangladesh. The combined flow of the Ganges, Jamuna (the downstream continuation of the Brahmaputra), and Meghna rivers flow to the Bay of Bengal, draining almost all the Himalayas, the most sediment-producing mountains in the world. A staggering total of one trillion cubic meters per year of water and one billion ton per year of sediment flow through the delta (Akter et al. 2016). As all deltas, the Bangladesh delta is low lying with 10 percent of the country less than 1 m above mean sea level and one third under tidal influence (Ali 1999). Eighty percent of the annual rainfall occurs during the monsoon season (June–September), resulting in very high river levels. In most years the peak volume exceeds the capacity of the river network, so flooding is a normal phenomenon. Aside from the high runoff volumes, several other factors contribute to flooding in Bangladesh, such as very low floodplain gradients and excessive siltation in the river beds (Rahman 1996). During many millennia of human occupation, Bangladeshi society has learned to live with floods (Novak 1994; Mishra 2001; Van Staveren et al. 2016).

At present, floods inundate 25 to 30 percent of the country in normal seasons. This normal flooding is highly desired by the rural population because these floods bring in new fertilizer to the paddy fields, in the form of silt, and new fish stock to the village ponds. However, extreme events occasionally result in significant loss of lives and cause serious adverse impacts on the national economy. The frequency of such disastrous events seems to be increasing. For example, in 1988 and 1998 about 61 and 68 percent, respectively, of the country flooded for a duration of between 15 and 45 days (Penning-Rowsell et al. 2013). This increased frequency should not automatically be taken as proof of climate change. Large upstream dams in India and reduction of the river capacity due to the building of flood control measures in Bangladesh have changed the behavior of the river (Thompson and Warburton 1985). At the same time, increased population density and economic investment have increased the risk of potential damages. However, the loss of land through river bank erosion during (normal and extreme) floods causes rural distress more than the occurrence of extreme floods, for which people have learned to cope. It should be noted, however, that rural distress is caused less by the occurrence of extreme floods, for which people have learned to cope, than by the loss of land through river bank erosion during (normal and extreme) floods. Riverbank erosion displaces an estimated 50,000–200,000 people in Bangladesh each year (BBS 2011).

While the northern part of Bangladesh is vulnerable to riverine flooding, the southern coastal part of Bangladesh is mainly prone to flooding from storm surges and cyclones (Tingsanchali et al. 2005). Devastating coastal flooding events from cyclones occurred in 1970 and 1991, causing a loss of 300,000 and 130,000 lives, respectively. In addition, like many other deltas, Bangladesh is experiencing compaction and/or isostatically induced subsidence, which is increasing the flooding effect (Goodbred and Kuehl 1999; Sarker et al. 2011, 2014). This situation causes a relative sea-level rise even before considering the projected climate change impacts on sea levels. Coastal areas also experience regular tidal flooding from June to September when the sea is in spate due to south-westerly monsoon winds. Due to the magnitude of flows and sediment loads, structural interventions can never eliminate the risk of flooding completely. The history of structural measures to reduce flood impacts is long (Van Staveren et al. 2016); however, their destruction rate is high and maintenance costly. Since the 1990s investment and expertise in non-structural flood risk management (including early warning, evacuation, rehabilitation, flood shelters) have significantly reduced the human impact of floods.

Despite recurrent, extensive flooding from rivers and the sea, Bangladesh is a very densely populated country with more than 150 million people, or approximately 1,000 people per km^2, and growing by approximately 2 million per year. Around 80 percent of the total population lives in the floodplain areas (BBS 2011). Agriculture is a mainstay of the economy of Bangladesh, both as a source of exports and of livelihoods. At the same time, every year the country is losing land and crops because of flooding and riverbank erosion.

The extremely poor people living in the char-lands (river islands) are the ones most exposed to and affected by these hazards (Penning-Rowsell et al. 2013). In most years many of these people flee temporarily to nearby higher areas in response to floods, many households have members who move to urban or other rural areas for seasonal work due to limited opportunities in the chars, and each year some households move permanently to nearby lands or further to urban areas due to the disappearance of their land through erosion. In urban areas they find employment as unskilled laborers on building sites, as rickshaw drivers, or, in fewer cases, in the booming garment industry. This largely temporary and seasonal migration contributes to the rapid growth of Bangladeshi cities, where an expanding urban economy does not reduce levels of social inequality. Floodplain development for construction purposes in cities in turn exacerbates flood risk conditions. While the Bangladesh national climate adaptation strategy (MoEF 2005, 2009) views urban migration as negative, migration also enables people to cushion shocks and improve their livelihood in entrepreneurial ways—being trapped back home may be the worse option. We discuss the intricacies of migration dynamics in Bangladesh and possible links with climate change in Section 5.4. First, we provide a context to the Bangladeshi debates by sketching the emergence of the current dominant global discourse on climate change and migration.

5.3 The Rise of Climate Change Migration as a Global Issue

The relationship between climate change and (international) migration was first considered in the mid-1980s by environmental activists and scientists. Based on crude impact assessments and assumptions that all inhabitants of affected areas would leave, they painted an apocalyptic picture of large numbers of climate refugees amassing at the borders of rich countries (e.g., Myers 1989; Kaplan 1994; for an overview, see Bettini 2013). Although these estimates still circulate, "[i]n the absence of empirically compiled evidence of the global historical scale of the phenomenon, such projections hardly qualify as sound scientific practice" (Moriniere and Hamza 2011). The end of the Cold War raised the hope of freeing up substantial resources (the peace dividend) for development which at that time used for defense (Hartmann 2010; Methmann and Oels 2015). An influential neo-Malthusian strand, however, predicted resource wars and violence induced by environmental refugees. The 1990s saw environmental security rise to prominence in the Clinton Administration, and get its second wind after Al Gore's Inconvenient Truth film made the headlines in the 2000s. This discourse also resonates with respect to climate migration to Dhaka, which people fear will bring Malthusian violence to its slums, negating the positive contribution of informal migrant networks in supporting food security during flood disasters (Keck and Edzold 2013).

The security discourse thus claims that a singular and significant relationship between climate change and migration is a threat to the existing world order. This discourse was initially used by activists of many backgrounds to argue that climate change should be taken seriously (Piguet 2013) but was quickly adopted in global and national climate change policy arenas. Over time the emphasis in the climate refugee discourse shifted from state security to the human security of migrating people (Biermann and Boas 2008), a humanitarian concern which, however, also echoes an imputed liability of the industrialized West for causing anthropogenic climate change, which then led to climate refugees (Mayer et al. 2013). A resilience discourse has superseded this reasoning in recent years. States, NGOs, and multilateral organizations have converged on the idea of resilience-as-adaptation (as expressed in the Sustainable Development Goals, the Sendai declaration on disaster risk reduction, and other initiatives), shifting responsibility to supposedly empowered individuals and portraying migration as an individual livelihood option (Bose 2016). These shifts could constitute three different discourses; however, the current global dominant discourse intermingles concerns from the three phases.[1]

Not surprisingly then, Moriniere and Hamza (2012) find several discourses in their Q-method study of the current global climate-migration debates. They distinguish these discourses by their problem-and-solution framing, the terminology they use, the assignment of responsibility, and (dis)agreement about the possibility to quantify what they label environmentally induced mobility. Each professional group prefers the discourse specific to its roles. For example, NGO employees tend to focus on human needs and rights and want to act, while employees from international organizations are generally more doubtful about interventions when drivers for migration are not clear and so argue for more research. The adherents of each of these four discourses on climate migration include academics—as noted, the proposition that climate change would lead to (mass) migration originated in the environmental sciences. Environmental sciences are still at the core of most of the research commissioned by (inter)national agencies and governments, although the research teams now include a strong representation of social scientists.[2] However, the focus tends to be on predictive (post)positivist, macro-scale quantitative research aimed at finding causal relationships, or at least predictions of future trends, possibly using scenarios.

---

1 Piguet (2013) labels the three phases "alarmist," "skeptical," and "pragmatic." Methmann and Oels (2015) ground their analysis in Foucauldian governmentality theory and see them as reflecting the subsequent governmentality rationales of sovereign power, liberal biopower, and advanced liberal government, respectively (see also footnote).

2 As an imperfect but illuminating illustration: a recent search for publications with "climate" and "migrant" or "migration" in their topic listed in Web of Science (WoS) yielded 17,392 records, of which the WoS classified approximately 75 percent in the environmental or technical sciences.

Curiously, the much more well-established domain of migration studies has only recently been re-discovered as potentially relevant for understanding climate migration. When it became recognised that migration is a more complex phenomenon than what had been assumed, this echoed what migration scholars had been saying for a long time, they were not consulted (Piguet 2013). Piguet (2013, 156) posits that "these insufficiencies in the research are attributable to the long eclipse of the environment within mainstream migration studies and its reappearance through the back door in the work of alarmist environmental scientists." This disjunction may be related to the fact that migration studies steered clear of considering the natural environment in looking for drivers for migration for fear of seeming deterministic (Castles 2011; Piguet 2013). Leading researchers, policy makers, and donors—who prefer scenario modeling and large-N household studies for country-level assessments of the future—continue to consider qualitative, constructivist, critical case studies that do not look to produce generalized statements as irrelevant.[3] However, rich case studies are becoming more common, an early example being the EU project "Environmental Change and Forced Migration Scenarios (EACH-FOR)." The project involved 23 case studies across all inhabited continents and used forced migration scenarios developed within the project (Warner 2011; Warner and Afifi 2014). In the tradition of the environmental sciences, EACH-FOR adopted a (semi)deterministic framework and did not endeavor to analyze the impact of scientific and policy discourses and power relations. The project had insufficient opportunity for comparing and contrasting the case studies to see if contextual and non-contextual aspects of climate change and migration could be extracted and verified (Upadhyay et al. 2015, 407).

In summary, most scholars currently agree that migration has many causes and that climatic variability is just one of several factors that explain migration (Foresight 2011; Gemenne and Baldwin 2013). Environment is one factor driving migration, interacting with many others, including power relations, class structures, economic inequalities, colonial legacies, cultural and religious specificities, political organization, and gender relations. It is doubtful that causal relations exist in the deterministic sense since the effect of the environment is not just due to objectively measurable physical changes, but (more importantly) dependent on subjective ways in which individuals and societies make sense of these changes—which cannot be captured in a deterministic model (Demeritt 2001; Piguet 2013). The relationship between the environment and migration is rarely direct, causal, or unidirectional, but is instead often indirect, contextual, and/or with feedback loops. However, to take action researchers have to eliminate

[3] This situation is not dissimilar to the dichotomy and status in the studies of water-human systems described in Wesselink et al. (2016).

some of the complexity, so that they consider only a few factors and relationships between factors. This method has political implications since this gives policymakers the opportunity to pick a politically convenient choice which is not necessarily incorrect but nevertheless biased (Upadhyay et al. 2015, 394).

Overall, "the primacy of climate change adaptation and mitigation efforts and planning has become so pronounced that one might argue that this is a new and pre-eminent form of development in the international arena [..] the 'climate change turn' in development work" (Bose 2016, 168), with Dutta and others (2011, 16) suggesting that "climate change is the new package through which the World Bank is reinventing itself."[4] It is exactly these politics of the climate-migration research and policy making that is a cause for concern. From our brief overview of the literature, a pattern emerges of a majority of studies that neglect underlying, long-term political drivers of assumptions behind the analysis (Felli and Castree 2012). This neglect, in turn, limits the range of policy responses to draw from, while providing a political excuse for not raising difficult questions of abuses of power, biased or inaccessible legal courts, and corruption (Upadhyay et al. 2015). Arnall and others (2014) show how the globally dominant discourse fits in the general landscape of global climate change politics. The effect of these politics boils down to naturalization and depoliticization: there is little regard for societal dimensions such as equity and unequal power relations, instead the focus is on safe issues such as natural phenomena and technologies. In this scheme, the presumed mode of governance tends to be technocratic and neoliberal and does not question existing power relations. This process in turn regularly gives rise to the appropriation of the discourse by vested interests to push through previously unaccepted unpopular policies. Arnall and others (2014, 100) conclude that "greater scrutiny of the discursive and political dimensions of mitigation and adaptation activities is required, with more attention being directed towards the policy consequences that governments and donors construct as a result of their framing and rendition of climate change issues. Such an approach will require constant questioning of the underlying epistemological and ethical assumptions underpinning and framing these debates [..], such as those relating to the ongoing discussion on global environmental change and migration."

[4] The related emergence of a "climate change research industry" is reflected in the number of peer-reviewed papers listed in the WoS. On 19/01/2017 the search combinations "climate" and "migration" or "migrant" yielded less than 10 records/year until the early 1990s, steadily increasing to around 100/year by 2000, to 1,000 by 2016. Adding other related search terms will increase these numbers significantly.

5.4 Climate Change and Migration in Bangladesh

With its precarious hydraulic and societal situation, "[i]mages of rising waters, flooded fields and displaced farmers in the region have become an iconic symbol deployed during debates on climate change both locally and globally. As a result, Bangladesh has emerged as a laboratory of sorts in which a series of national-level strategic plans, projects, programmes, trust funds and financing schemes are being designed and tested in partnership with international donors and development agencies, all built around the idea of climate change and resilience" (Bose 2016, 168). Each of these plans, projects, programs, trust funds, and financing schemes endeavors to be (seen to be) evidence based, so in addition to the significant investments in developing policies and in actual schemes to adapt to the effects of climate change, including helping local populations adapt, organizations spend large amounts of money on funding local or foreign researchers to collect this evidence. In this section, we first give an overview of such studies and then discuss whether and how their outcomes are used in policy and projects.

Prior to the rise of climate change as a global issue, researchers published a significant amount of research on the relationship between migration and environmental factors in Bangladesh, such as riverbank erosion, flooding, and cyclones. By extension, this literature can also help policy makers to assess the possible impacts of climate change on migration. In these early publications, authors assert that people are used to "living with floods" (Haque 1988; Haque and Zaman 1993; Novak 1994; Rashid 2000; Mishra 2001) and that "[t]he arrogance of scientists and engineers in controlling the nature has created more problems than solved them. [..] Why don't we think of converting the flood water into a resource instead of dealing with it as a problem" (Mishra 2001). At the same time, it is the poor who suffer most: "Most people displaced by riverbank erosion [..] are already poor and disempowered before being uprooted by the shifting local channel of the Jamuna River. Thereafter, a significant proportion faces even bleaker prospects. Nevertheless, those who have resettled locally have done much to maximize their chances within the limits of powerful natural, economic, social, and political constraints. [..] they have been resilient, resourceful and innovative in ways that their conventional representation as passive victims of natural and social forces outside their control would not suggest" (Indra 2000). In this and other literature, researchers do not consider normal flooding in itself a problem. Instead, it is a necessity to recharge groundwater, maintain crop yields (although studies debate its contribution to soil fertility and mechanisms whereby crop yields are achieved; see Brammer 2004), and restock and maintain floodplain fisheries. Major riverine floods have in the past caused widespread losses with people living in the chars and close to the main rivers the worst affected, and this risk remains. However, it is the riverbank erosion associated with peak flows that is a main source of longer lasting rural impoverishment and migration.

Like the academic publications on the generic relationship between climate change and migration discussion in Section 5.3, the number of publications on climate change and migration in Bangladesh has exploded since 2010. When climate change first entered the agenda, most of the research in Bangladesh focused on changes to the physical systems and impacts on agricultural production (Mirza 2002; Begum et al. 2006; Walsham 2010) with one or two exceptions (some of the chapters in Warrick and Ahmad 1996 and Brouwer et al. 2007). At the current date (early 2017) climate change is prominent in social science publications on migration in Bangladesh.[5] We briefly present a representative selection of these studies. Most of the studies are based on qualitative case study research, often in two or three locations, sometimes using statistical methods to analyze the findings (Joarder and Miller 2013). They focus on issues such as:

- Factors affecting whether environmental migration is temporary or permanent (Joarder and Miller 2013)
- Population movement as a last resort in response to climate-related hazards (Penning-Rowsell et al. 2013)
- Vulnerability to rainfall variability and food insecurity in northern Bangladesh and its effects on migration (Etzold et al. 2014)
- The comparative analysis of coping in two different vulnerable areas (Yasmin and Ahmed 2013)
- A behavioral model for climate-related migration in rural Bangladesh (Martin et al. 2014)
- The role of microfinance in household livelihood adaptation (Fenton et al. 2017)
- Migrating to tackle climate variability and change in coastal fishing communities (Islam et al. 2014)

As indicated by these topics, although these publications list climate change in the keywords, they do not generally emphasize or even analyze climate change as a driver of migration; climate change figures as possibly exacerbating existing drivers for and mechanisms of migration. The reason can, of course, be because they didn't find climate change to be a driver of migration. After all it is difficult and contested to determine actual climate changes over the recent past in Bangladesh (it is easier to discern at higher latitudes

[5] On 18/01/2017 a WoS search for "climate" or "climatic" and "migration" or "migrant" and Bangladesh in the title or topic yielded 98 articles. This search does not cover all relevant search terms such as combinations with "refugee" and "environment" instead of either "migration/migrant" or "climate/climatic."

and altitudes). An exception is Ahsan and others (2014, 1) who are adamant that "climate change is a driver of social change," which according to them results in "an explosion of climate migrants" and "is particularly apparent through the creation of a new social community of 'climate migrants'." However, their research does not support these statements, which is based on interviews with urban migrants to discover why they relocated; contrary to the assertions made in the introduction of the paper, climate change does not figure among the reasons given.

Generally, researchers agree that "in Bangladesh [..] environmental degradation has been shown to interact in a complex way with other factors in shaping livelihood and migration strategies. It intervenes as a contingent additional burden in the context of pre-existing cultures of mobility where the population affected are not passive victims but purposive actors" (Piguet 2013, 157) with "different areas having different knowledge of migration opportunities, with differing assessments being made regarding the desirability of moving" (Findlay and Geddes 2011, 150). This acknowledgement of complex interactions between push and pull factors is also what Stojanov and others (2014) found in their research on Bangladeshi expert discourses on climate and migration. However, in the literature pull factors, such as employment opportunities, receive much less attention than push factors, that is, rural issues such as livelihoods. This reflects the more general rural development bias in Bangladesh that neglects analyses of urban poverty and development (Banks et al. 2011).

In spite of the rich case descriptions (which this brief summary does not do justice to), we find that most studies on Bangladesh that adopt the keywords climate change and migration, while not being deterministic in the strict sense, naturalize the topic by depoliticization: they suggest that displacement is an inevitable consequence of changing weather conditions and/or population dynamics. An exception is Choudhury and Haque (2016) who investigate at local scales how the interplay of institutions, community power structures, and human agency shape adaptive capacity of wetland communities. They let the voices of marginalized rural dwellers speak: "We are more scared of the power elites than the floods." The recent overview by Roy and others (2016), "Bangladesh confronts climate change. Keeping our heads above water," also includes analysis of power and politics. More numerous analyses of this type are found in published research with keywords such as displacement and land grab. For example, Feldman and Geisler (2012, 791) describe how some locations of riverine and coastal erosion and sedimentation are "contested sites ripe for power plays that uproot small producers on their rich alluvial soils" resulting in "in-situ displacement, where people may remain in place or experience a prolonged multi-stage process of removal"—which is much less visible than ex-situ displacement where people move away from their homes, communities, and livelihoods. Baqee (1998) describes similar processes of power and social relations affecting the settling of chars. Adnan (2013) relates local land grabs and peasant resistance in the coastal zone to national and international developments, such as neoliberal globalization; he shows how the establishment of a shrimp zone for export production led to systematic eviction of the poor, backed by state power. Critical research is not a recent development, as evidenced by Zaman's work. Zaman (1991) puts the rise of exploitative relations between landlords and peasants in floodplains in an historical perspective and demonstrates how interconnections between the rural landlords and political power at the national level, as well as a weak state system, allow local deployment of violence and power abuse to remain unpunished.

However, policy documents and policy-oriented research from the Government of Bangladesh and from international donors fail to consider such studies on abuses of power and land grab as causes for migration. The policy discourse in such policy documents remains highly simplistic: the impact of climate change is unquestioned, overrules all other push factors, and presents reason for alarm. This contradicts the ample evidence of complexity, multi-causality, and trends that pre-date the entrance of climate change as a cause for concern. Key national-level climate change and development initiatives are the National Adaptation Plan of Action (MoEF 2005) and the 10-year Climate Change Strategy and Action Plan (BCCSAP) (MoEF 2009). Organizations have set up several funds to finance projects to implement these plans, such as a US$200 million Bangladesh Climate Change Trust Fund and a donor-funded US$120 million Bangladesh Climate Resilient Fund. To coordinate responses and collect evidence, the government of Bangladesh set up the Climate Change Unit housed in the Ministry of Environment and Forests. The government also included climate change as a key pillar in its 2016–2020 5-year economic plan. Government documents identify two migration-related issues as a special concern: the slums of the capital Dhaka as destination area and source areas along the Jamuna river and along the coast. The NAPA only mentions migration as a negative livelihood impact, with an increase in urban migration as result. However, the BCCSAP makes numerous references to the potential migration impacts of climate change and advocates "support to adversely impacted populations through capacity building for re-settlement and rehabilitation in their new environment." However, according to Walsham (2010, 33–34) in his overview, "Environment, Climate Change & Migration in Bangladesh", for the International Organization for Migration, "despite these encouraging signs within more recent policy on climate change, it is fair to say that at present migration issues are not effectively mainstreamed with environmental, disaster management, or climate change policy."

This policy discourse resonates with the climate refugees discourse discussed in Section 5.3 and exemplifies the new development mantra pointed out by Bose (2016) (see Section 5.3). Our own ongoing research is revealing how actors use the simplistic discourse to raise the alarm by exaggeration and dramatization, while at the same time they recognize the complexity of migration and climate change issues. We found several examples of policy studies where the simplistic discourse appears in the introduction and/or summary, while the body of the work discusses multiple mechanisms and a nuanced evaluation of benefits and problems (Shamsuddoha and Chowdhury 2009; Mehedi 2010; CDMP II 2014). The main purpose of simplification appears to be to attract international attention and funding, sometimes including assignment of blame. According to one study, "Bangladesh demands compensations for climate change-related loss and damage from the industrialized countries" (Vriens et al. 2017).

Overall, it is, therefore, not clear how research results inform the government policies and allocation of funding. Many of the peer-reviewed publications discussed here are outcomes of commissioned work funded by international organizations such as the International Organization for Migration, the World Bank, or externally funded research programs (e.g., Climate Development Knowledge Network funded by DFID/ESRC; DECCMA and HIAWARE funded by IDRC and DFID). One may expect that these funding agencies and the government are eagerly awaiting the results and will make attempts to turn them into action. However, the peer-reviewed publications as well as a lot of the grey literature indicate that every location has its own specificities and each segment of the population has its own needs and obstacles. This means that generalization is hardly possible: there are no one size fits all solutions—if it would be possible to identify a problem needing a solution without contestation to start with. As long as governments and aid organizations focus on the national level and on technocratic, high-cost, and de-politicized solutions that do not challenge business-as-usual, one cannot expect that the funds allocated to climate change will have much real impact on impoverishment, disempowerment, and migration.

5.5 An Outlook: Prudent Governance of Migration

The conundrum that we sketched was analyzed by Nicholson (2014), who convincingly argues that an issue is prone to political capture if there are no generalizable causal relationships, as is the case for migration, since in these cases evidence-based policy is not possible. In the case of Bangladesh, the political capture takes the shape of a simplistic security-based discourse that favors alarmist narratives, a development that Mayer and others (2013) also observed. These tendencies result in a lack of effective governance strategies to address the issue due to the complex and multi-causal character of environment-related migration. "It is difficult in the first place to neatly attribute specific human activities to changing environmental and climatic conditions; as such changes typically reflect a range of interactions between human activities and environmental settings. Floods, for instance, do not occur simply because of heavy rain, but rather because such precipitation outstrips water management systems. Likewise, salinity intrusion in southern Bangladesh follows from sea-level rise as well as damage to embankments by cyclones, but also from large-scale shrimp farming. Even as such relationships between societies and natural environments become better understood, the ramifications of environmental changes and abrupt events still intrinsically depend upon the social contexts in which they play out" (Mayer et al. 2013, 178).

How then to proceed? We suggest that strategies and policies must go beyond science-based designs or the usual top-down policy-making processes if the goal is to move forward in supporting equitable adaptation to climate change. We draw on Aristotle's notion of *phronesis* (translated as prudence or practical wisdom) as the guiding principle. In ethics, a prudential judgment is where one must weigh the circumstances to determine the correct action. Prudence deals with the ability to deliberate about contingent, variable, or indeterminate matters, when the right course of action is not self-evident from facts. Aristotle adds that the right course of action depends upon the details of a specific situation, rather than merely from applying a law.[6] This interpretation, of course, is a very fitting description of the issue of migration in Bangladesh. This, of course, is a very fitting description of the issue of migration in Bangladesh. The question 'what is good' then becomes as much if not more appropriate as 'what is true', and more relevant. Answering the first question means taking sides – either for elites and major donor-funded programmes, or for the poor and activist bottom-up initiatives that enable 'practical action' that is purposeful and concerted (cf. Ezrahi 1980).

References

Adnan, S. 2013. Land grabs and primitive accumulation in deltaic Bangladesh: Interactions between neoliberal globalization, state interventions, power relations and peasant resistance. *The Journal of Peasant Studies* 40(1):87–128.

Afifi, T., and K. Warner. 2008. *The Impact of Environmental Degradation on Migration Flows across Countries*. Working Paper No. 5. Bonn, Germany: United Nations University Institute for Environment and Human Security (UNU-EHS).

Ahsan, R., S. Karuppanan, and J. Kellett. 2014. Climate induced migration: Lessons from Bangladesh. *The International Journal of Climate Change: Impact and Responses* 5(2). http://ijc.cgpublisher.com/product/pub.185/prod.230/m.2

Akter, J., M. H. Sarker, I. Popescu, and D. Roelvink. 2016. Evolution of the Bengal Delta and its prevailing processes. *Journal of Coastal Research* 32(5):1212–1226.

[6] This explanation of *phronesis* draws on Wikipedia entries for prudence and *phronesis*.

Alam, M. 2004. Adverse impacts of climate change on development of Bangladesh: Integrating adaptation into policies and activities. *Report Prepared Under Capacity Strengthening of Least Developed Countries (LDCs) for Adaptation to Climate Change* (CLACC). Dhaka, Bangladesh: BCAS.

Ali, A. 1999. Climate change impacts and adaptation assessment in Bangladesh. *Climate Research* 12(2–3):109–116.

Arnall, A., U. Kothari, and I. Kelman. 2014. Introduction to politics of climate change: Discourses of policy and practice in developing countries. *The Geographical Journal* 180(2):98–101.

Auerbach, L. W., S. L. Goodbred Jr, D. R. Mondal et al. 2015. Flood risk of natural and embanked landscapes on the Ganges–Brahmaputra tidal delta plain. *Nature Climate Change* 5:153–157.

Bangladesh Bureau of Statistics. 2011. *Population Census of Bangladesh.* Dhaka, Bangladesh: Bangladesh Bureau of Statistics.

Banks, N., M. Roy, and D. Hulme. 2011. Neglecting the urban poor in Bangladesh: Research, policy and action in the context of climate change. *Environment and Urbanization* 23(2):487–502.

Baqee, A. 1998. *Peopling in the Land of Allah Jaane. Power, Peopling and Environment: The Case of Char-Lands of Bangladesh.* Dhaka, Bangladesh: University Press.

Begum, S., A. M. K. Uddin, R. Ernst, and N. Haque. 2006. *Climate Change and Bangladesh Annotated Bibliography.* Comprehensive Disaster Management Programme, Department of Environment. Dhaka, Bangladesh: Climate Change Cell.

Bettini, G. 2013. Climate Barbarians at the Gate. A critique of apocalyptic narratives on 'climate refugees'. *Geoforum* 45:63–72.

Biermann, F., and I. Boas. 2008. Protecting climate refugees: The case for a global protocol. *Environment Magazine.* http://www.environmentmagazine.org/Archives/Back%20Issues/November-December%202008/Biermann-Boas-full.html

Bose, P. S. 2016. Vulnerabilities and displacements: Adaptation and mitigation to climate change as a new development mantra. *Area* 48(2):168–175.

Brammer, H. 2004. *Can Bangladesh be Protected from Floods?* Dhaka, Bangladesh: University Press.

Brouwer, R., S. Akter, L. Brander, and E. Haque. 2007. Socioeconomic vulnerability and adaptation to environmental risk: A case study of climate change and flooding in Bangladesh. *Risk Analysis* 27(2):313–326.

Castles, S. 2011. Concluding remarks on the climate change-migration nexus. In *Migration and Climate Change,* E. Piguet, A. Pecoud, and P. De Guchteneire (Eds.), pp. 415–427. Cambridge, UK: Cambridge University Press.

CDMP II. 2014. *Trend and Impact Analysis of Internal Displacement due to the Impacts of Disaster and Climate Change. Comprehensive Disaster Management Programme (CDMP II), Ministry of Disaster Management and Relief.* Dhaka, Bangladesh: Department of Disaster Management.

Choudhury, M. I., and C. E. Haque. 2016. 'We are more scared of the power elites than the floods': Adaptive capacity and resilience of wetland community to flash flood disasters in Bangladesh. *International Journal of Disaster Risk Reduction* 19:145–158.

Demeritt, D. 2001. The construction of global warming and the politics of science. *Annals of the Association of American Geographers* 91(2):307–337.

Dutta, S., A. Munshi, P. Khanna, and J. Athialy. 2011. *World Bank, Climate Finance and Bangladesh: A Briefing Note.* New Delhi, India: Bank Information Center.

Etzold, B., A. U. Ahmed, S. R. Hassan, and S. Neelormi. 2014. Clouds gather in the sky, but no rain falls. Vulnerability to rainfall variability and food insecurity in Northern Bangladesh and its effects on migration. *Climate and Development* 6(1):18–27.

Ezrahi, Y. 1980. Utopian and pragmatic rationalism: The political context of scientific advice. *Minerva* 18(1):111–131.

Feldman, S., and C. Geisler. 2012. Land expropriation and displacement in Bangladesh. *The Journal of Peasant Studies* 39(3–4):971–993.

Felli, R., and N. Castree. 2012. Neo-liberalising adaptation to environmental change: Foresight or foreclosure? *Environment and Planning A* 44(1):1–4.

Fenton, A., J. Paavola, and A. Tallontire. 2017. The role of microfinance in household livelihood adaptation in Satkhira District, Southwest Bangladesh. *World Development* 92:192–202.

Findlay, A., and A. Geddes. 2011. Critical view on the relationship between climate change and migration: Some insights from the experience of Bangladesh. In *Migration and Climate Change*, E. Piguet, A. Pécoud, and P. De Guchteneire (Eds.), pp. 138–159. Cambridge, UK: Cambridge University Press.

Foresight. 2011. *Migration and Global Environmental Change*. London, UK: The Government Office for Science.

Gemenne, F., and A. Baldwin. 2013. The paradoxes of climate change and migration. In *World Social Science Report: Changing Global Environments*, ISSC (Ed.), pp. 265–268. Paris, France: UNESCO.

Goodbred, S. L. Jr, and S. A. Kuehl. 1999. Holocene and modern sediment budgets for the Ganges-Brahmaputra river system: Evidence for high stand dispersal of floodplain, shelf, and deep-sea depocenters. *Geology* 27(6):559–562.

Hancock, G. 1989. *Lords of Poverty*. London, UK: Macmillan.

Haque, C. E. 1988. Human adjustments to river bank erosion hazard in the Jamuna floodplain, Bangladesh. *Human Ecology* 16(4):421–437.

Haque, C. E., and M. Q. Zaman. 1993. Human responses to riverine hazards in Bangladesh: A proposal for sustainable floodplain development. *World Development* 21(1):93–10.

Hartmann, B. 2010. Rethinking climate refugees and climate conflict: Rhetoric, reality and the politics of policy discourse. *Journal of International Development* 22:233–246.

Indra, D. 2000. Not just dis-placed and poor: How environmentally forced migrants in rural Bangladesh recreate space and place under trying conditions. In *Rethinking Refuge and Displacement: Selected Papers of Refugees and Immigrants*, Vol. VIII, E. M. Gozdziak and D. Shandy (Eds.), pp. 163–191. Washington, DC: American Anthropological Association.

Islam, Md. M., S. Sallu, K. Hubacek, and J. Paavola. 2014. Migrating to tackle climate variability and change? Insights from coastal fishing communities in Bangladesh. *Climatic Change* 124:733–746.

Joarder, Md. A. M., and P. W. Miller. 2013. Factors affecting whether environmental migration is temporary or permanent: Evidence from Bangladesh. *Global Environmental Change* 23:1511–1524.

Kaplan, R. D. 1994. *The Coming Anarchy: How Scarcity, Crime, Overpopulation, Tribalism, and Disease Are Rapidly Destroying the Social Fabric of Our Planet*. Boston, MA: Atlantic Monthly.

Keck, M., and B. Etzold. 2013. Resilience refused: Wasted potentials for improving food security in Dhaka. *Erdkunde* 67(1):75–91.

Martin, M., M. Billah, T. Siddiqui et al. 2014. Climate-related migration in rural Bangladesh: A behavioural model. *Population and Environment* 36:85–110.

Mayer, B., I. Boas, J. Ewing et al. 2013. Governing environmentally-related migration in Bangladesh: Responsibilities, security and the causality problem. *Asian and Pacific Migration Journal* 22(2):177–198.

Mehedi, H. 2010. *Climate Induced Displacement Case Study of Cyclone Aila in the Southwest Coastal Region of Bangladesh*. Khulna, Bangladesh: Coastal Livelihood and Environmental Action Network.

Methmann, C., and A. Oels. 2015. From 'fearing' to 'empowering' climate refugees: Governing climate-induced migration in the name of resilience. *Security Dialogue* 46(1):51–68.

Mirza, M. M. Q. 2002. Global warming and changes in the probability of occurrence of floods in Bangladesh and implications. *Global Environmental Change* 12(2):127–138.

Mishra, D. K. 2001. Living with floods: People's perspective. *Economic and Political Weekly* 36(29):2756–2761.

MoEF. 2005. *National Adaptation Programme of Action (NAPA)*. Final Report, Ministry of Environment and Forest (MoEF), Government of the People's Republic of Bangladesh. Dhaka, Bangladesh: Ministry of Environment and Forest.

MoEF. 2009. *Bangladesh Climate Change Strategy and Action Plan (BCCSAP)*. Action Plan, Ministry of Environment and Forests (MoEF), Government of the People's Republic of Bangladesh. Dhaka, Bangladesh: Ministry of Environment and Forests.

Moriniere, L. C. E., and M. Hamza. 2012. Environment and mobility: A view from four discourses. *AMBIO* 41(8):795–807.

Myers, N. 1989. Environment and security. *Foreign Policy* 74(1):23–41.

Nicholson, C. T. M. 2014. Climate change and the politics of causal reasoning: The case of climate change and migration. *The Geographical Journal* 180(2):151–160.

Novak, J. J. 1994. *Bangladesh: Reflections on the Water*. Dhaka, Bangladesh: University Press.

Penning-Rowsell, E., P. Sultana, and P. Thompson. 2013. The 'last resort'? Population movement in response to climate-related hazards in Bangladesh. *Environmental Science & Policy* 27(1):44–59.

Piguet, E. 2013. From "primitive migration" to "climate refugees": The curious fate of the natural environment in migration studies. *Annals of the Association of American Geographers* 103(1):148–162.

Powell, M., and D. Seddon. 1997. NGOs & the development industry. *Review of African Political Economy* 24(71):3–10.

Rahman, A. 1996. Peoples' perception and response to floodings: The Bangladesh experience. *Journal of Contingencies and Crisis Management* 4(4):198–207.

Rashid, S. F. 2000. The urban poor in Dhaka city: Their struggles and coping strategies during the floods of 1998. *Disasters* 24(3):240–253.

Roy, M., J. Hanlon, and D. Hulme. 2016. *Bangladesh Confronts Climate Change. Keeping Our Heads above Water*. London, UK: Anthem Press.

Salauddin, M., and M. Ashikuzzaman. 2011. Nature and extent of population displacement due to climate change-triggered disasters in the south-western coastal region of Bangladesh. *Management of Environmental Quality* 22(5):620–631.

Sarker, M. H., J. Akter, M. R. Ferdous, and F. Noor. 2011. Sediment dispersal processes and management in coping with climate change in the Meghna Estuary, Bangladesh. In *Proceedings of the ICCE Workshop Held at Hyderabad, India*, September 2009. IAHS Publication 349, pp. 203–217.

Sarker, M. H., C. R. Thorne, M. N. Aktar, and M. R. Ferdous. 2014. Morpho-dynamics of the Brahmaputra–Jamuna River, Bangladesh. *Geomorphology* 215:45–59.

Shamsuddoha, M., and R. K. Chowdhury. 2009. *Climate Change Induced Forced Migrants: In Need of Dignified Recognition Under a New Protocol*. Dhaka, Bangladesh: Equity and Justice Working Group.

Stojanov, R., B. Duží, and I. Kelman. 2014. Local experts' perceptions of climate change and migration in Bangladesh. In *Environment, Migration and Adaptation – Evidence and Politics of Climate Change in Bangladesh*, M. Bishawjit and B. Etzold (Eds.), pp. 213–229. Dhaka, Bangladesh: AHDPH.

Thompson, M., and M. Warburton. 1985. Decision making under contradictory certainties: How to save the Himalayas when you can't find out what's wrong with them. *Journal of Applied Systems Analysis* 12:3–34.

Tingsanchali, T., and M. F. Karim. 2005. Flood hazard and risk analysis in the southwest region of Bangladesh. *Hydrological Processes* 19(10):2055–2069.

Upadhyay, H., I. Kelman, G. J. Lingaraj et al. 2015. Conceptualizing and contextualizing research and policy for links between climate change and migration. *International Journal of Climate Change Strategies and Management* 7(3):394–417.

Van Staveren, M., J. Warner, and S. A. Khan. 2016. Bringing in the tides. From closing down to opening up delta polders via Tidal River Management in the southwest delta of Bangladesh. *Water Policy* 19:147–164.

Vriens, W., L. W. Auerbach, S. L. Goodbred Jr et al. 2017. Framing migration in the context of climate change in Bangladesh. MSc Research report. http://hydro-social-deltas.un-ihe.org/sites/hydro-social-deltas.un-ihe.org/files/vriens_final_report_cc-m_discourses.pdf

Walsham, M. 2010. *Assessing the Evidence: Environment, Climate Change and Migration in Bangladesh. Evidence Dialogue Strategies*. Dhaka, Bangladesh: International Organization for Migration.

Warner, K. 2011. Environmental change and migration: Methodological considerations from ground-breaking global survey. *Population and Environment* 33(1):3–27.

Warner, K., and T. Afifi. 2014. Where the rain falls: Evidence from 8 countries on how vulnerable households use migration to manage the risk of rainfall variability and food insecurity. *Climate and Development* 6(1):1–17.

Warrick, R. A., and Q. K. Ahmad. 1996. *The Implications of Climate and Sea-Level Change for Bangladesh*. Dordrecht, the Netherlands: Kluwer Academic Publishers.

Yasmin, T., and K. M. Ahmed. 2013. The comparative analysis of coping in two different vulnerable areas in Bangladesh. *International Journal of Scientific & Technology Research* 2(8):26–38.

Zaman, M. Q. 1991. Social structure and process in char land settlement in the Brahmaputra-Jamuna floodplain. *Man* (*New Series*) 26(4):673–690.

6

Changing Social Capital in the Mountains and the Implications for Adaptation Interventions: An Exploratory Analysis with Case Studies from the Hindu Kush Himalaya

Arabinda Mishra

CONTENTS

6.1 The Context 90
6.2 Social Capital and Adaptation to Climate Change 92
6.3 Methodology 93
6.4 Trust and Reciprocity in Times of Diminishing Resources Availability and Productivity 95
6.4.1 Case 1: Drying Springs and Farmers Managed Irrigation Systems 96
6.4.2 Case 2: Traditional Pastoralism in High Altitude Rangelands 97
6.5 Depleted Traditional Networks Due to Increased Migration 98
6.5.1 Case 3: Feminization of Agriculture and Emergence of Women Bonding Social Capital 98
6.6 Displaced Norms and Customary Practice Due to Government Programmes 100
6.6.1 Case 4: Transforming Shifting Cultivation 100
6.7 Trust, Norms, and Networks Activated in Post-Disaster Settings 101
6.7.1 Case 5: Post-Disaster Recovery of Mountain Communities 101
6.8 Discussion and Conclusion 103
Acknowledgment 104
References 105

6.1 The Context

Multiple forces of change, including climate change, are acting on natural resources-dependent hill and mountain communities in the Hindu Kush Himalaya (HKH) region.[1] The significant growth in the use of phone-based and internet-based communications has led to growing exposure of local socio-economic systems to the outside world, and in the process has fueled aspirations among people even in remote locations. Human mobility, primarily in the form of out-migration of male youth from the hills and mountains to the urban centers in valleys and plains, is a growing phenomenon. The expanding urban centers have placed ever-increasing demands on the natural resource systems. The region has experienced significant economic growth; however, the distribution of development gains across regions and populations remains grossly uneven.

Additional to these forces of change in the HKH, climate change is introducing new risks and vulnerabilities in the HKH region that are not yet fully understood. But it is likely that climate change will cause an increase in the frequency and magnitude of extreme weather events such as high intense rainfall leading to flash floods and landslides (Shrestha et al. 2015). These events would most likely adversely affect various productive sectors in the HKH countries, particularly the natural resource systems providing livelihoods for poor and marginal communities (Ahmed and Suphachalasai 2014). Adding to the uncertainty is the way multiple stressors like climate change and globalization may interact in the mountain context to produce differential vulnerabilities (O'Brien et al. 2004).

Remote mountain areas limit the reach of externally provided support services (including the extension services by government line agencies) and force local communities to rely on their own ingenuity and collective action to adapt to adversity arising out of change (Jodha 1992). Characteristics of mountain communities are their strong solidarity, collective efforts based on mutual trust and reciprocity, and institutionalized norms that define social behavior (some illustrative examples are in the 1999 International Center for Integrated Mountain Development (ICIMOD) literature). The literature identifies these characteristics as sources of social capital in a community. As defined by Putnam (1993) in one of the early definitions of the concept, "social capital … refers to features of social organization, such as trust, norms, and networks that can improve the efficiency of society by facilitating coordinated actions." However, mountain communities is a pluralistic concept and there is a huge diversity characterized by ethnicity, language, and culture.

[1] The Hindu Kush Himalayan (HKH) region extends 3,500 kilometers over all or part of eight countries from Afghanistan in the west to Myanmar in the east (http://www.icimod.org/?q=1137).

Moreover, the forces of change, directly and indirectly, possibly influence social capital in local mountain communities. The popular discourse on this front is that social capital is eroding. Thus, male out-migration results in a withering away of traditional farmer networks for knowledge-sharing and resource-sharing in mountain agriculture (Mishra et al. 2017a). Formal governance arrangements at the local level provide little scope for traditional community institutions—taken as a reflection of social capital—to retain their relevance (ICIMOD 1999) and so on. On the other hand, there are observations from post-disaster contexts (e.g., the 2015 earthquake in Nepal) suggesting that even in heterogeneous communities the social capital can arise in novel forms and effectively contribute to the post-disaster recovery of the community (Mishra et al. 2017b). The proposition that emerges from such diverse observations is that perhaps it is not that there is always an erosion of social capital because of the forces of change, but that quite likely there is a modification of the form and nature of social capital in many cases and in ways not yet fully understood.

Social capital possibly facilitates community level collective efforts to build its adaptive capacity (Adger 2001; Aldrich et al. 2016; Harrison et al. 2016) to deal with adverse situations arising from disasters, disruptions, and resource scarcity. Such collective response is possible when communities can self-organize and self-regulate through formal or informal institutional arrangements. Jodha (1992) identified the inherent ability of mountain communities to adapt to change as a mountain specificity, but he also acknowledged that with changes related to population, market, and state, many of the human adaptation mechanisms were losing their efficacy and feasibility. With increasing concerns regarding the impacts of climate change, the urgency to act on adaptation intervention is growing for national governments in the HKH region. Researchers and policymakers are looking at mechanisms to harness social memories and experiential learning, which are likely to help build the adaptive capacity of communities through anticipation and learning (Gunderson 2010). Given these circumstances, it is pertinent to examine the implication of changing social capital for community level adaptation responses that require some degree of social bonding and collective effort.

This chapter is an exploratory analysis to gain an understanding about the influences acting on the sources of social capital in rural mountain communities in the HKH region. The proposition is that multiple forces are interacting with each other to produce a process of simultaneous decay and emergence of social capital in the communities, and that this has implications for the planned adaptation interventions. The exploration is through five case studies that are representative of the major influences on rural communities in the HKH region.

In terms of structure, the chapter begins with a brief review of the literature that relates social capital with adaptation to climate change. The next section provides a description of the case study-based methodology used for finding some answers to the key research question: how is social capital in mountain communities influenced by phenomena such as disasters, out-migration, increasing resource scarcity and degradation, and development programs by the government? The next three sections present the illustrative case studies. In the final section there is a discussion of the implications of changing social capital in the HKH for planned adaptation actions followed by the conclusion.

6.2 Social Capital and Adaptation to Climate Change

The many definitions of social capital point to the same set of sources—trust, norms, and networks—contained in the definitions by Putnam (1993). The outcome of social capital primarily manifests itself in the form of collective action based on formal or informal institutional arrangements. We now associate Institutions themselves with social capital as a representation of its stock in a community or society (Ostrom 1992). One form of investing in social capital is to make possible the evolution of effective institutions for resource governance and benefit-sharing from resource use.

The contribution of social capital for community adaptation to climate change is in terms of a collective effort that arises from bonding, bridging, and linking types of social relations. Bonding social capital brings together people with similar objectives, bridging enables groups with different views to come together, and linking is the ability of groups to engage with external agencies that have power and resources (Pretty 2003). "Adaptation to climate change is inevitably local" (Agrawal 2008, 1), and the most effective adaptation responses "will be based on bottom-up, local level components, facilitated by social capital that are also coordinated across society" (Aldrich et al. 2016, 10). Adger (2001) asserts that "civil society institutions can play an important role in coping with the impacts of climate variability and change and can be encouraged through appropriate interventions."

Following are some examples of social capital's contribution to adaptation that draw primarily from the comprehensive review of Aldrich and others (2016). In disaster settings bonding social capital can contribute to resource pooling (Agrawal 2008), influence decisions to evacuate or stay (Horney et al. 2012), and provide immediate post-disaster support (Aldrich and Sawada 2015). Bridging and linking social capital are important in the post-disaster situation to access extra-local resources, political support, and quicker bureaucratic response—all of which can contribute to accelerated recovery (Aldrich 2016). Informal solidarity networks constituted or strengthened after climate-related disasters are also manifestations of social capital contributing to recovery and adaptation (OECD [Organization for Economic Cooperation and Development] no date). From the HKH region, Banerjee and others (2017) provides the example of flood-affected households borrowing money from sources (friends, relatives and so on) within their social network which is illustrative of their capacity for risk pooling. Similarly, the study by Islam and Walkerden (2014) of cyclone-hit villages in Bangladesh suggests that while both bonding and bridging networks provide great support in post-cyclone recovery, bridging support is weaker if everybody in the larger community is a victim of the disaster, and support can wither over time.

Many traditional risk-sharing mechanisms based on social capital, such as asset pooling and kinship networks, may not work well for climate risks because climate risks can be all encompassing in a society and can surprise because of their unexpected consequences (OECD no date). There is growing advocacy for adaptive governance to deal with increased uncertainty and risk associated with future climate change impacts. Most of the criteria necessary for a successful transformation toward adaptive governance are a function of social capital and include collaboration, co-management, participation, trust, leadership, social memory, and the formation of actor groups (Folke et al. 2005). To link diverse sets of actors, facilitate interactions across knowledge systems, and lower transaction costs of collaboration, the adaptive governance approach proposes bridging organizations. These are formal institutions which provide an arena for knowledge co-production, trust building, sense making, learning, vertical and horizontal collaboration, and conflict resolution (Pahl-Wostl et al. 2013; Kowalski and Jenkins 2015).

6.3 Methodology

This study looks at social capital as structural property (Portes 1998, 21) of rural mountain communities. In its community level application, social capital is "simultaneously cause and effect" (Portes 1998, 19) and, to avoid the inevitable tautological conclusions from analysis, the recommendation from Woolcock (1998, p. 185) is to "focus primarily on its sources rather than its consequences." Accordingly, this study focuses on (1) mutual trust within communities along with reciprocity between communities, (2) norms underlying customary law and practice for natural resource management (specifically land resource), and (3) informal networks among farmers as the sources of social capital. Among the multiple forces of change acting on these sources, the study chooses to illustrate the phenomena of resource depletion and degradation, male out-migration, large scale government interventions, and natural hazards that are typical to the HKH region as being mountain-specific in their nature.

On the consequences side of the analytical framework, we move a step away from concepts associated with the definitions of social capital and, instead, introduce the notion of institutional dependence of mountain communities on which there is some baseline information available from surveys by ICIMOD. From 2009 to 2013, ICIMOD conducted an extensive participatory assessment exercise, covering 90 villages across 15 districts of Bhutan, India, and Nepal and spanning an altitudinal range varying from 50 to 3,500 mean sea level. The poverty and vulnerability assessment survey provides data on, among other things, the institutional dependence of mountain communities on four categories of services: (1) to address basic needs, (2) to access productive resources, (3) to access extension services, and (4) to get social support (ICIMOD 2014). This quantitative exercise provides us with a baseline pattern of institutional dependence of mountain communities as summarized in the following list:

- *Categories*: The assessment categorizes institutions into civic (informal and formal), public, and private institutions. The informal civic institutions include social (family, clan, peer groups) and customary institutions (traditional bodies). Under the formal civic institutions, there are SHGs (Self Help Groups), user groups, non-governmental organizations (NGOs), and religious bodies. The assessment categorizes government agencies and statutory bodies under public institutions, and banks and market institutions under private institutions.
- *Basic needs*: An appraisal of the results suggests that communities in Uttarakhand and northeast India are primarily dependent on informal civic institutions, such as friends, neighbours, relatives, and clan, when they require help in meeting basic needs. In sharp contrast, communities in Bhutan turn to public institutions for provision of these services. In Nepal, private institutions provide these services—primarily the local shopkeeper or moneylenders.
- *Access to resources*: For access to resources, formal civic institutions are important for communities in Nepal, Uttarakhand, and northeast India. Institutions such as the affinity groups—SHGs and their federations—are extremely important when communities need access to micro-credit. Religious bodies such as the church play an important role in northeast India, helping to provide access to resources. In the case of Uttarakhand and Bhutan, communities access credit from the private institutions—banks in Bhutan and shopkeepers and moneylenders in Uttarakhand.
- *Access to technical and extension services*: Public institutions largely meet this need in all the study areas. However, for access to development related support, such as government schemes, income generating activities, and infrastructure, the NGOs and donor projects become extremely important. In Bhutan, government agencies provide such services and no other institutions have any relevance.
- *Social support*: For the purposes of social support, such as provision of moral and spiritual support at times of need, the formal civic institutions, in particular the affinity groups, become very important for communities in Uttarakhand and Nepal. In northeast India, traditional women, youth, or church groups take up this role.

TABLE 6.1

Major Influences on Sources of Social Capital in the HKH and the Consequent Implications for Adaptation

Phenomena (Region-Wide, Mountain-Specific)	Case Number	Study Area	Source/s of Social Capital Influenced	Consequences on Institutional Dependence	Implications for Adaptation Planning
Diminishing resources availability and productivity	1 & 2	Nepal, India	Mutual trust (and reciprocity)	Shift from community to private	Conflict management
Increasing male out-migration	3	Nepal, Pakistan	Networks	Dense networks of male farmers replacement by emergent bonding among women farmers	Investment in women-friendly extension services
Growing government interventions	4	India, Myanmar	Norms	Customary institutions becoming irrelevant	Mechanisms to avoid social exclusion; creating bridge organizations
Increasing incidences of disasters	5	Bangladesh, Myanmar, Nepal	All three – trust, norms, and networks	Greater reliance on civic institutions in the immediate aftermath of crisis	Strengthening state-society linkages; promoting social learning

The five case studies that follow illustrate whether the changes in sources of social capital are in any way likely to influence the previous list of baseline patterns of institutional dependence. Obviously, the number of case studies is too small to judge the significance of changes, if any, to the baseline pattern. However, it sets the stage for further studies in this direction. Moreover, from the anticipated changes in the pattern of institutional dependence we hope to draw the possible implications for adaptation interventions in the region. Table 6.1 presents a snapshot of the present study based on the methodology previously discussed.

6.4 Trust and Reciprocity in Times of Diminishing Resources Availability and Productivity

There is already a significant evidence base to establish that increasing climate variability and climate change is already impacting on many natural resource systems in the HKH region that are closely linked to livelihoods.

In this study we focus specifically on mountain springs and rangelands. In the following two case studies we get a glimpse of the consequences of increasing resource scarcity and decreasing resource productivity on social capital in local communities that have been managing these systems through well-developed institutional arrangements.

6.4.1 Case 1: Drying Springs and Farmers Managed Irrigation Systems

Springs richly endow Nepal's mid-hills. As ICIMOD's research has shown, a few tens of square kilometers around a hill water tower can have over a hundred large and small, seasonal and permanent springs (Sharma et al. 2016). These springs are valuable resources for rural mountain communities, often playing a critical role for their very survival. Recently springs are drying up in the hills, leading to significant loss of local livelihoods and out-migration to urban areas. Many of Nepal's farmer managed irrigation systems (FMIS) have springs as their water sources, and with the drying up of springs these irrigation systems are under pressure.

The Panchakanya irrigation system is a small (600 ha) gravity irrigation system fed by a network of springs known as *panchanadi*. This is a 200-year-old system that a formal water users committee has managed for the last 20 years. For the past 10–15 years, famers in the area have reported significant decrease in rainfall amount, change in rainfall pattern, and increase in temperature. According to famers, winter rainfall has almost ceased to occur, and the dry period has prolonged. As a result, the amount of irrigation water in the canal has decreased. Farmers also believe that another reason for the decrease in water in the canal is the deep boring practices in the upstream settlements. Apart from this, there is continuous sedimentation that occurs every year during the monsoon around the *panchanadi* sources, which depletes the water coming out of the ground.

Decreasing water availability has triggered several changes in the institutional arrangements linked to the Panchakanya FMIS that indicate an erosion of social capital in the farming community. Over the years, there has been growing conflict between the users and users committee, resulting in polarization between the two parties. The committee has been unable to distribute water to the tail users nor has it been able to maintain and renovate the canals. So far, the committee has been unable to even collect the *paanipot* (water tax) from all its users regularly. The committee is slowly losing the trust of its users. Above all, the lack of awareness about climate change and its impacts is making the committee aloof to the potential use of water efficient technologies and water efficient crops. The result is a growing irrelevance of the existing approach to governance of the system. The weakening of the community institution has led to several consequences that support the narrative which sees new forms of associations and networks replacing existing arrangements. In 2007–2008, farmers from one branch separated from the Panchkanya irrigation system and opted for the community and private deep bore wells that came with government co-financing support. Many small land-holding famers, as a coping strategy, have been forming groups and hiring electric and diesel pumps to extract groundwater from big land-holding farmers.

But there are new vulnerabilities as well. The trend towards privatization of irrigation has made farming costly for small landholders, which has led to other members in these families out-migrating to seek work in nearby factories. The consequence for landless laborers within the command area is even more severe. These laborers have been working for generations as agricultural workers. However, with many landowners either moving away from agriculture altogether, or hiring contractors from other parts of the country to get cheaper and mechanized agricultural help, the landless agricultural laborers have become more vulnerable because they have no form of livelihood options. In these communities, men of the households have started working in factories and as construction laborers, but women are unemployed. The vulnerability is highest in families where there are no men and women head the households.

6.4.2 Case 2: Traditional Pastoralism in High Altitude Rangelands

Rangelands form the largest land use system (60 percent) in the HKH region (Miller and Craig 1996) and contribute to the livelihood of 25–30 million pastoralists and agro-pastoralists (Ning et al. 2013). Rangeland herders are among the poorest and most vulnerable communities in the HKH. Surveys of pastoral communities provide evidence of this vulnerability in terms of worsening food deficits within households (Ning et al. 2013). Because of climate change, researchers expect the availability and productivity of rangeland resources, including the existence of water points, to decrease with marked consequences for mountain livelihoods.

Historically, local communities managed grazing lands in the Himalayas as common property resources under customary frameworks (Oli et al. 2013). The communities intricately intermeshed migratory pastoralism with other livelihood activities, such as trade and agriculture, and local level arrangements sustained the community and included customary reciprocal ties with neighbouring communities and their rangelands across the national borders (Pandey et al. 2017).

The traditional practices and migration patterns of pastoralism have decreased in recent years. An ICIMOD study (Pandey et al. 2017) conducted during 2014–2015 in select districts of India (Pithoragarh district of Uttarakhand state) and Nepal (districts Baitadi, Darchula, Bajhang, and Humla) attributes the decrease to several reasons such as closure of international borders, friction with statutory institutions, switch to alternative livelihood options, and so on—none of which really capture the direct impact of climate change on the resource system. But the breakdown of cross-border migratory pastoralism certainly implies an erosion of the previous transboundary regime of reciprocity.

In many parts of the HKH there is evidence emerging to suggest that because of climate change treelines are shifting upward, with potential effects on alpine rangeland. In an ongoing study by ICIMOD's partner in the high-altitude region of the Kanchenjunga Conservation Area (RHF 2017), many local respondents stated that the treeline has been shifting upward and thus the rangeland is getting narrower. The number of herders has decreased in recent years and there has been a sharp reduction in the population of livestock and transhumance in the study villages. Most of the local people have shifted to non-agricultural practices (such as business, medicinal plant trade) from agricultural farming, livestock rearing, and transhumance movement. All this change is possibly adversely affecting the local knowledge system and social memory around community-led traditional rangeland management.

6.5 Depleted Traditional Networks Due to Increased Migration

For people in the mountains of HKH, circular labor migration has traditionally been a livelihood strategy. Typically, the young male members in a household migrate temporarily (with periods ranging from months to multiple years) to adjoining towns or even out of the home country and send back money. Traditionally such type of human mobility contributed to monetization of traditional barter and reciprocity based rural economies, increased exposure to the outside world, flow of new skills to the local economic system, and networking among migrant workers outside of own village and community. The following case study looks at the phenomenon in more recent times and specifically at the social impact it has had on women members left behind in the local communities.

6.5.1 Case 3: Feminization of Agriculture and Emergence of Women Bonding Social Capital

In recent times the trend of labor out-migration has accelerated because of the growth in small and medium-sized towns, increased aspirations among the youth, traditional farm-based occupations turning non-remunerative, and lack of economic opportunities at the local level. Along with the circular labor migration there is also the phenomenon of entire households migrating permanently to urban centers in the valleys and plains to obtain better educational and health facilities, and the urban amenities. In its extreme form entire households have permanently migrated out of villages resulting in depopulated villages (ghost villages), complete collapse of traditional institutions, and depletion of larger community networks (Pathak et al. 2017).

ICIMOD's analysis of its own survey data finds very high rate of labor migration in the mountain areas. Particularly, more than 40 percent of households in the Upper Indus and Koshi river basins in Pakistan and Nepal have sent domestic and international out-migrants to diversify their livelihoods (ICIMOD 2012). At this scale of male labor migration, traditional responsibilities of agriculture, natural resource management, and disaster preparedness have shifted to the women family members left behind (Mishra et al. 2017a). Women must perform labor-heavy and physically taxing farming-related tasks previously assigned to men, such as land preparation, seedbed preparation, woodcutting, and threshing. Climate change impacts are additional stressors for women because of the changes in the availability of water, fodder, and fuel wood.

However, despite women's changed roles and responsibilities, they still must deal with the same institutions, at local and national levels, geared towards economic decision making by men (Gioli et al. 2014). In a major study on agriculture extension in Nepal's mid-hills, government-employed agriculture extension service providers are mostly male and this often discourages women farmers from interacting with and benefiting from the available extension support (Mishra et al. 2017a). In any case, the reach of government extension system in the hills and mountains is in most parts highly inadequate owing to geographical remoteness and a lack of connectivity. Under such circumstances farmers' networks play a major role in the spread of improved technology, knowledge and practices. Studies have observed such a role by the farmers' networks elsewhere in the developing world (Yishay and Mobarak 2013). These male-dominated networks are becoming ineffective because the young males are increasingly out-migrating and women farmers find it difficult to take part in these networks.

During 2013–2017, ICIMOD conducted action research in three separate sites in the HKH. The sites are Lakhimpur district of Assam in India, the Hunza and Nagar districts of Upper Indus sub-basin in Pakistan, and Udaypur district in Nepal. About 700 women members from migrant-sending households belonging to the three sites participated in the action research. The action research interventions are in the form of customized trainings and village level extension services that aim to increase the women's awareness and ability to process information about financial services, flood preparedness, and livelihood diversification, and accordingly plan the utilization of their own resources (including financial remittances from out-migrated family members).

The action research formed a women's group in each study village. Being in a peer group seems to encourage the women to be more pro-active in the community, and to instill confidence in themselves to take up leadership roles. Many among the group members are sharing the knowledge received from the trainings with their relatives and neighbors. These women are encouraging others to consider target-based savings, prepare for floods, and find new ways to diversify household income. In terms of social capital, a sense of bonding has developed among the women of the groups that encourages them to share household and village level issues with each other and find a solution. In the Udaypur site this bonding has matured to a stage in which there is active networking among women in the community to spread knowledge about new farm practice (e.g., use of biochar as fertilizer) and market trends.

6.6 Displaced Norms and Customary Practice Due to Government Programmes

The literature on common property resources management provides ample evidence of social capital contributing to successful collective action by local communities (Ostrom 1990; Sorensen 2000; Pretty 2003). Many indigenous and ethnic minority groups in the HKH region have historically been practicing shifting cultivation and continue to do so. In the traditional system of shifting cultivation, groups manage land as common property and norms ensure some form of access to the resource by all members. Restrictions to shifting cultivation and the promotion of settled systems have alienated vulnerable members of communities who previously had access to land under customarily managed shifting cultivation systems.

6.6.1 Case 4: Transforming Shifting Cultivation

Ukhrul district in Manipur has one of the larger tribes of the Naga indigenous people, the Tangkhul Naga, which still maintains a traditional lifestyle defined by age-old conventions and traditions and pursues shifting cultivation and rain-fed terrace cultivation as a primary occupation. The local self-governing unit is the *hanga* (village council) that has clan elders headed by the hereditary chief. The *hanga* is responsible for land and resources management and regulation of the rules and norms for upholding customary laws and practices.

> According to the Tangkhul Naga customary laws, land is sacrosanct. The community belongs to the land and not land to the community. Land tenure and land use system in the Tangkhul emanates from this concept where the ownership of land belongs to the future generation. Therefore, when a family inherits land it is the tenurial custodian role and responsibility for the land and not the ownership of the land. As customary practice a family must undergo consultation at different levels (relative, clan, village, and tribe) to sell land (Vashum 2016).

The government is introducing transformation through the North Eastern Region Community Resource Management Project (NERCORMP). Among other things, the project has interventions designed for strengthening village governance; fallow management and expansion of community conservation area; *jhum* plot consolidation; promotion of home gardens; and livelihood diversification. The apprehension of the community is that in the long run the transformation process will lead to land use as private property. Along with this there is the growing market integration that is adding pressure to change land use form.

Management of land as a common resource is not the only aspect of traditional shifting cultivation that promotes inclusivity in the community. Traditional *jhum* is also about sharing of seeds, joint labor, and is a practice that is more a way of life than simply a livelihood. In contrast, the transformation process introduces the dependence on market for seeds, encourages the use of wage or contract labor, and excludes the landless from having access to land. It is inevitable that this process will irreplaceably affect the traditional way of life and the bonding in the community that goes with it.

A similar story comes from shifting cultivators of Myanmar. Under the present *lopil* system of customary communal tenure, everyone in the community, including the landless, may have access to land in the shifting cultivation plots. The villagers must agree upon the selection of individual plots in coordination with the village authority. The same is true of water and forest resources. Under transformation to the terrace system or orchards, however, this system breaks down and exclusion emerges.

6.7 Trust, Norms, and Networks Activated in Post-Disaster Settings

The HKH is one of the most disaster-prone regions in the world. Mountain people are resilient because of their ability to cope with frequent and multiple hazards, such as landslides, flash floods, avalanches, and forest fires. But even among the mountain people post-disaster recovery can vary from community to community. Social capital is one of the factors that is generally associated with early recovery from disasters.

6.7.1 Case 5: Post-Disaster Recovery of Mountain Communities

During 2015–2016, ICIMOD conducted a multi-country, multi-hazard study motivated by the desire to identify factor combinations contributing to community resilience in terms of quick and better recovery from disasters (Mishra et al. 2017b). The study identified 14 key contributing factors from literature including social capital. They collected the data from 77 communities in 3 countries (Nepal, Bangladesh, and Myanmar) through focus group discussions. The study followed a Qualitative Comparative Analysis (QCA) methodology[2] to identify multiple factor combinations contributing to post-disaster recovery of the communities.

Generally, the presence of strong social capital had positively contributed to early recovery. Communities that recovered early were quickly able to self-organize and respond collectively to external rescue and relief operations. These communities had the least conflicts while sharing external aid. A strong sense of connectedness encouraged the people to share their grief with each other. Locals said they shared their remaining food, accommodation, and money with each other. They also mentioned that the feeling of being together at difficult times helped them heal psychologically and reduced their fear and anxiety. Adherence to social norms helped people maintain order and function in an organized manner at the time of chaos. Similarly, we observed the presence of strong social capital in communities that were the first ones able to shift to new settlements. Moving to a new settlement is a long and painstaking process as it involves finding new land, obtaining a large amount of funds, and negotiating with donors and authorities. Strong social bonding united people in their demands, approach, and efforts which was effective in persuading authorities and donors.

The more interesting result from the study is the identification of multiple factor combinations contributing to early recovery from disasters. These factor combinations were not only context dependent, but also individual factors complementing and substituting each other. Thus, social capital (indicators being, for example, instances of collective action, sharing of community assets, social codes of conduct, and so on) combined with institutions (as evidenced from the presence of traditional societies, user group associations, SHGs, cooperatives, and so on) substituted for the absence of social homogeneity; on the other hand, social capital complemented institutions. Thus, in case of landslide-affected communities in Chin State of Myanmar, despite social heterogeneity (multiple denominations of Christianity), there was strong evidence of collective action, self-organization by setting up of committees, sharing of food, and so on. In one of the study townships, for example, post-landslide the residents started rebuilding houses communally: the entire village would assist in building one house, then move to the next. Families borrowed most of the money to meet the construction costs from relatives.

[2] QCA is a formalized technique based on set theory to capture causal relations between multiple factor combinations and a specific outcome (Ragin 1987).

6.8 Discussion and Conclusion

The five case studies presented in this chapter help in building a narrative on social capital in the context of climate change in the HKH region. There are multiple drivers including climate change interacting among themselves and directly or indirectly inducing change in local institutions that have traditionally managed natural resource systems. Climate change impacts on resource availability and productivity either triggers or accentuates the change (ranging from a withering away to opening up), which in turn affects local livelihoods. Out-migration, mostly of the male youth, introduces a gender dimension to the institutional change. Concurrent to the influences originating from the resources side, there is often the crowding-out role by development programs introduced by external agencies, which leads to marginalization of traditional institutional arrangements and may result in social exclusion. At the same time, during times of crises coming together of factors bring community members together to make use of their social capital for early and better recovery.

An erosion of social capital is certainly bad news for adaptation governance by public authorities. As the FMIS case (Case 1) illustrates, with the weakening of resource managing institutions there is the risk of conflict among users and the risk of elite capture of the institutions themselves. Further, as social capital decreases in a community the transaction costs of cooperation on adaptation actions increase. In case of transboundary resource systems like rangelands (Case 2) such costs can reach levels that make cooperation impossible. The influence of other factors (such as closing of international borders) costs the loss of a valuable form of social capital (transboundary social networks and reciprocity).

The other negative consequence of reduced social capital is the exclusion of marginalized groups from resource systems. The case of shifting cultivation shows (Case 4) that traditional institutional arrangements are often better at achieving inclusion, with social capital often offering social safety nets in times of crises. In the design of adaptation interventions, it would therefore make sense to follow an approach that mixes the good elements of traditional practices with that of modern science-led actions. Customary law and practice that is good for adoption would need a certain degree of legitimacy from the modern political system so that they are not crowded out by the modern forms of governance arrangements and so that they remain a source of social capital for the local communities. It is in such cases that bridge organizations can play a relevant role.

The case of feminization of agriculture in the HKH (Case 3) brings in the gender dimension to our discussion of social capital and adaptation. What the case study seeks to establish is that women need their institutional space in local communities. Out-migration leads to breakdown of the male-dominated farmer-to-farmer networks through which traditionally dissemination on agriculture practices and technology occurs. These networks also connect to the government system for accessing agri-extension services. Effective adaptation by women farmers requires corresponding networking and connectivity arrangements for flow of information. Further, it is likely that men and women will have different ways of mobilizing social capital, so there is a need to build the capacity among women farmers to take advantage of the institutional arrangements.

Finally, the case of post-disaster recovery (Case 5) has an important message. Social capital does play a significant role in enabling communities, even in case of non-homogenous ones, to overcome the hardships of a crisis and recover in a better manner. Investing in adaptation and resilience building requires that we know the context-specific combination of factors, especially the ones that complement social capital (e.g., institutions) or can act as substitutes (e.g., access to NGOs) in its absence.

Studies already accept social capital as one of the five forms of capital (the other four being natural, physical, human, and financial) that communities mobilize in the pursuit of their livelihood objectives. Communities require investment and state support to build or strengthen the social capital most appropriate to a specific context. Institutions play a major role in defining the adaptive capacity of communities and public authorities should keep in mind the whole range of formal and informal civic institutions as important assets that provide critical support services to vulnerable households. Climate change interacting with other socio-economic drivers in the HKH region influences social capital in mountain communities in diverse ways. The dependence of communities on institutions for services is accordingly getting modified. The planning for adaptation action by public authorities would need to take this process of change in social capital to effectively harness the synergy arising out of state-society linkage and develop interventions tailored to the context.

Acknowledgment

The views and interpretations in this publication are those of the author and are not necessarily attributable to ICIMOD. The author is grateful to several ICIMOD colleagues: Rucha Ghate, Soumyadeep Banerjee, Suman Bisht, Sabarnee Tuladhar, and Anjal Prakash, and also Krity Shrestha of Practical Action-Nepal, for helping in preparing the case studies.

References

Adger, W. N. 2001. Social capital and climate change. Tyndall Centre Working Paper No. 8. Norwich, UK: University of East Anglia.

Agrawal, A. 2008. The role of local institutions in adaptation to climate change. Social Development Working Papers, Paper No. 118. *Paper Prepared for the World Bank Conference*, Washington, DC.

Ahmed, M., and S. Suphachalasai. 2014. *Assessing the Costs of Climate Change and Adaptation in South Asia*. Mandaluyong, Philippines: Asian Development Bank.

Aldrich, D. P., and Y. Sawada. 2015. The physical and social determinants of mortality in the 3.11 tsunami. *Social Science & Medicine* 124:66–75.

Aldrich, D. P., C. Page, and C. J. Paul. 2016. Social capital and climate change adaptation, Oxford Research Encyclopedia. *Climate Science*. doi:10.1093/acrefore/9780190228620.013.342

Banerjee, S., D. Kniveton, R. Black, S. Bisht, P. Das, and S. Tuladhar. 2017. Does financial remittance build household level adaptive capacity? A case study of the flood affected households of Upper Assam in India. KNOMAD Working Paper 18. https://www.knomad.org/publication/do-financial-remittances-build-household-level-adaptive-capacity-case-study-flood

Folke, C., T. Hahn, P. Olsson, and J. Norberg. 2005. Adaptive governance of social-ecological systems. *Annual Review of Environment and Resources* 30:441–473.

Gioli, G., T. Khan, S. Bisht, and J. Scheffran. 2014. Migration as an adaptation strategy and its gendered implications: A case study from the Upper Indus Basin. *Mountain Research and Development* 34(3):255–265.

Gunderson, L. 2010. Ecological and human community resilience in response to natural disasters. *Ecology and Society* 15(2):18. http://www.ecologyandsociety.org/vol15/iss2/art18/

Harrison, J. L., C. A. Montgomery, and J. C. Bliss. 2016. Beyond the monolith: The role of bonding, bridging, and linking social capital in the cycle of adaptive capacity. *Society & Natural Resources* 29(5):525–539. doi:10.1080/08941920.2015.1103389

Horney, J. A., P. D. M. MacDonald, M. Van Willigen, and J. S. Kaufman. 2012. The importance of effect measure modification when using demographic variables to predict evacuation. *Risk, Hazards & Crisis in Public Policy* 3(1):1–19.

ICIMOD. 1999. Social institutions, local governance, and social capital: Foundations of social development and poverty reduction. *Issues in Mountain Development* 99/1. Kathmandu, Nepal: ICIMOD. http://lib.icimod.org/record/17131/files/IMD99-1.pdf

ICIMOD. 2012. Himalayan Climate Change Adaptation Programme (HICAP). Poverty and vulnerability assessment in the Eastern Brahmaputra (India), Koshi (Nepal), Upper Indus (Pakistan), and Salween-Mekong (China) river basins, 2011–2012 [data file]. Kathmandu, Nepal: ICIMOD.

ICIMOD. 2014. Livelihoods and ecosystem services in the Himalayas: Enhancing adaptation capacity and resilience of the poor to climate and socio-economic changes. IFAD Grant Completion Report, August 2014. Kathmandu, Nepal: ICIMOD.

Islam, R., and G. Walkerden. 2014. How bonding and bridging networks contribute to disaster resilience and recovery on the Bangladeshi coast. *International Journal of Disaster Risk Reduction* 10:281–291. doi:10.1016/j.ijdrr.2014.09.016

Jodha, N. S. 1992. Mountain perspective and sustainability: A framework for development strategies. In *Sustainable Mountain Agriculture,* N. S. Jodha, M. Banskota, and T. Partap (Eds.). New Delhi, India: Oxford and IBH Publishing.

Kowalski, A. A., and L. D. Jenkins. 2015. The role of bridging organizations in environmental management: Examining social networks in working groups. *Ecology and Society* 20(2):16.

Miller, D. J., and S. R. Craig (Eds.). 1996. *Rangelands and Pastoral Development in the Hindu Kush Himalayas.* Kathmandu, Nepal: ICIMOD.

Mishra, A., N. K. Agrawal, G. Thorup, S. Bisht, I. C. P. Lelkanger, and N. Gupta. 2017a. *Strengthening Women's Roles as Risk and Resource Managers at the Frontline of Climate Change, Adaptation Solution Brief 1.* Kathmandu, Nepal: ICIMOD. http://lib.icimod.org/record/32584/files/icimodASB.pdf

Mishra, A., R. Ghate, A. Maharjan, J. Gurung, G. Pathak, and A. Upraity. 2017b. Building ex ante resilience of disaster-exposed mountain communities: Drawing insights from the Nepal earthquake recovery. *International Journal of Disaster Risk Reduction* 22:167–178. doi:10.1016/j.ijdrr.2017.03.008

Ning, W., G. S. Rawat, S. Joshi, M. Ismail, and E. Sharma. 2013. *High Altitude Rangelands and Their Interfaces in the Hindu Kush Himalayas.* Kathmandu, Nepal: ICIMOD.

O'Brien, K., R. Leichenko, U. Kelkar et al. 2004. Mapping vulnerability to multiple stressors: Climate change and economic globalization in India. *Global Environmental Change* 14(4):303–313.

OECD. n.d. Poverty and Climate Change, Part 2, Adaptation lessons from past experience. https://www.oecd.org/env/cc/4494574.pdf

Oli, K. P., L. Zhandui, R. S. Rawal, R. P. Chaudhary, S. Peili, and R. Zomer. 2013. The role of traditional knowledge and customary arrangements in conservation: Trans-border landscape approaches in the Kailash Sacred Landscape of China, India and Nepal. Chapter 3 in *The Right to Responsibility: Revisiting and Engaging Development, Conservation, and the Law in Asia,* H. Jonas, H. Jonas, and S. M. Subramanian (Eds.). Tokyo, Japan: UNU Press.

Ostrom, E. 1990. *Governing the Commons: The Evolution of Institutions for Collective Action.* Cambridge, UK: Cambridge University Press.

Ostrom, E. 1992. *Crafting Institutions for Self-governing Irrigation Systems.* San Francisco, CA: Institute for Contemporary Studies Press.

Pahl-Wostl, C., G. Becker, C. Knieper, and J. Sendzimir. 2013. How multilevel societal learning process facilitate transformative change: A comparative case study analysis on flood management. *Ecology and Society* 18(4):58.

Pandey, A., N. Pradhan, S. Chaudhari, and R. Ghate. 2017. Withering traditional institutions? An institutional analysis of the decline of migratory pastoralism in the rangelands of the Kailash Sacred Landscape, western Himalayas. *Environmental Sociology* 3(1):87–100.

Pathak, S., L. Pant, and A. Maharjan. 2017. De-population trends, patterns and effects in Uttarakhand, India—A gateway to Kailash Mansarovar. ICIMOD Working Paper 2017/22. Kathmandu, Nepal: ICIMOD.

Portes, A. (1998). Social capital: Its origins and applications in modern sociology. *Annual Review of Sociology* 24:1–24.

Pretty, J. 2003. Social capital and collective management of resources. *Science* 302(December): 1912–1914.

Putnam, R. 1993. *Making Democracy Work*. Princeton, NJ: Princeton University Press.

Ragin, C. 1987. *The Comparative Method: Moving Beyond Qualitative and Quantitative Strategies*. London, UK: University of California Press.

RHF. 2017. Livelihood strategy under climate change stress: assessing climate vulnerability and preparing adaptation framework with Kanchenjunga Community in Nepal. A Project Progress Report, Resources Himalaya Foundation, May 2017, Kathmandu, Nepal.

Sharma, B., S. Nepal, D. Gyawali, G. S. Pokharel, S. M. Wahid, A. Mukherji, S. Acharya, and A. B. Shrestha. 2016. Springs, storage towers, and water conservation in the midhills of Nepal. ICIMOD Working Paper 2016/3. Kathmandu, Nepal: Nepal Water Conservation Foundation and ICIMOD.

Shrestha, A. B., N. K. Agrawal, B. Alfthan, S. R. Bajracharya, J. Marechal, and B. van Oort, Eds. 2015. *The Himalayan Climate and Water Atlas: Impact of climate change on water resources in five of Asia's major river basins*. Kathmandu, Nepal: ICIMOD, Grid-Arendal, and CICERO.

Sorensen, C. 2000. Social capital and rural development: A discussion on issues. Social Capital Initiative Working Paper No. 10. The World Bank, Social Development Department. http://worldbank.org/socialdevelopment

Vashum, T. 2016. Managing transformation in shifting cultivation: Lessons from Ukhrul district, Manipur. *Presentation Made in Workshop on Understanding Transformations in Shifting Cultivation*, Imphal, India, September, 2016.

Woolcock, M. 1998. Social capital and economic development: Toward a theoretical synthesis and policy framework. *Theory and Society* 27:151–208.

Yishay, A. B., and A. M. Mobarak. 2013. Communicating with farmers through social networks. Yale University economic growth center discussion paper no. 1030; Yale Economics Department Working Paper No. 121. https://economics.yale.edu/sites/default/files/benyishay_and_mobarak_sept_2013.pdf

7

Enhancing Adaptive Capacity through Education: A Case Study of Rural Mountain Communities, Sikkim, Eastern Himalaya India

Bhupen Mili, Suparana Katyaini, and Anamika Barua

CONTENTS

7.1 Introduction 109
7.2 Study Area and Methodological Approach 112
  7.2.1 Background of the Study Area 112
  7.2.2 Selection of Study Sites 114
  7.2.3 Data Collection Methods and Techniques 114
7.3 Data Analysis 117
  7.3.1 Multidimensional Poverty Assessment Tool 117
7.4 Results and Discussion 119
  7.4.1 Status of Education in East District, Sikkim 120
  7.4.2 Caste Differentiation in Education 122
  7.4.3 Gender Differentiation in Education 126
7.5 Conclusion 130
Acknowledgments 131
References 131

7.1 Introduction

The scientific community has established climate change as a real phenomenon (Intergovernmental Panel on Climate Change [IPCC] 2007). Although the basic science of climate change is simple, the causes and likely impacts of climate change on human beings are highly complex (Adger and Kelly 1999; United Nations Framework Convention on Climate Change [UNFCCC] 2007; Agrawal and Perrin 2008; Norris et al. 2008; Paavola 2008). The impacts of climate change are not uniform across space and time and vary within the same region, communities, and among individuals due to the various levels of vulnerability[1] of different ecosystems, economic sectors, and social groups to climate change (O'Brien and Leichenko 2000). This variation is especially true for the Eastern Himalayan region as studies have shown that the region is highly vulnerable due to high topographical variation, multiple hazards, and a complex interaction of social, ecological, political, and economic factors shaped by multiple ethnic and socio-economic groups (Shrestha et al. 1999; McDowell et al. 2013; Sapkota et al. 2016). Warming in the Eastern Himalaya is rapid and exceeds the global average of 0.74°C over the last 100 years (IPCC 2014; Pandey and Bardsley 2015) and threatens high-altitude ecosystems (Shrestha et al. 1999; Sharma et al. 2009). The climate changes are likely to affect the natural resources such as rivers and glaciers of the region, intensify natural hazards, and adversely impact the natural resources dependent livelihoods, thereby affecting the overall wellbeing of people (Sharma et al. 2009). Further, low education, limited livelihood options, poor communication, poor access to modern services, and inequitable access to productive resources burden the region, thereby reducing the communities' capacity to cope with climate-related natural disasters (Gentle and Maraseni 2012). Although traditional institutions as well as indigenous knowledge in the past played a significant role by making the mountain communities less vulnerable to uncertainties posed by global change (Barua et al. 2014), in recent times, climate change has accelerated, and the traditional social institutions and indigenous knowledge systems have started to erode. For example, there has been a gradual loss of ecosystem and biodiversity-related knowledge, traditional agricultural practices, and the loss of language which is the only mode of intergenerational transmission of knowledge (Singh et al. 2009). This loss has posed a serious threat to the wellbeing and livelihood of the poor mountain communities of Eastern Himalaya.

Adaptation to climate change has taken a prominent place in government agenda, particularly in the mountainous regions. However, for a successful adaptation strategy, it is crucial to understand the context-specific factors that increase or constrain adaptive capacity. Studies have developed many adaptive capacity frameworks to assess community's capacity to adapt, particularly in the context of the mountainous regions. These frameworks consider access to economic resources, technology, information, skills, farm assets, non-farm assets, education, healthcare services, gender and social equality, and so on (IPCC 2001; Smith et al. 2003; Jones et al. 2010; Dixit et al. 2012) as important factors determining a community's capacity to adapt. These factors are context specific; hence, the existing socio-economic and political context influences the community's ability to adapt to climate

[1] Vulnerability here is defined as the degree to which a system is susceptible to, and is unable to cope with the adverse effects of climate change (IPCC, 2007a).

change and their ability to respond and recover from climate change impacts (IPCC 2007; Adger et al. 2009).

There are many reasons for making investments in education evidenced by the abundant climate change literature showing that better-educated people tend to have higher adaptive capacity (Adger et al. 2004; Wamsler et al. 2012; Land and Hummel 2013; Striessing et al. 2013; IPCC 2014; Muttarak and Lutz 2014). Researchers have further agreed that the most effective long-term defense against the adverse impacts of climate change is strengthening the factors associated with enhancing human capital. Education can strengthen human capital because it is linked with improving health, eradicating extreme poverty, and reducing population growth (Wamsler et al. 2012; Land and Hummel 2013; Striessing et al. 2013; Muttarak and Lutz 2014). Further, studies show that an educated society is more empowered and more adapted to recover from the climate change-related natural disasters (Brink 2010; Wamsler et al. 2012). Climate change literature (Brink 2010; Magnan 2010; Wamsler et al. 2012; Waston 2014) states that education enables a person to make well-informed decisions and empowers communities by enlarging the set of opportunities available to them, thereby contributing significantly to building their adaptive capacity. Hence, studies consider education as a crucial indicator of adaptive capacity (Adger et al. 2004a; Brink 2010; Magna 2010).

The crucial indicators of education as determinants of adaptive capacity are quality (teacher-pupil ratio in a class), availability (school: primary and below, middle, secondary, higher secondary; college: bachelors and above), and access (distance to the educational institute) (Unterhalter et al. 2007). Socio-cultural, economic, and political context further influence accessing and utilizing the interventions (Marcus and Harper 2014). For example, despite girls having access to education, they face barriers in attending school due to gender-based family practices, which limits their achievements (United Nations International Children's Fund [UNICEF] 2003). Similarly, a study conducted by Unni (2009) found that benefits from government projects in India, particularly on education have not resulted in achievement of desired benefits from interventions for marginalized and vulnerable sections of the population in terms of gender, religious, and social classification (UNICEF 2003; Unni 2009). Further, Unterhalter (2009, 219) mentioned, "the same level of resources may be quite inadequate for children who are shy, hungry, with poor concentration, always sit at the back of the class and talk a minority language." Hence, while school as a resource is available or accessible, it does not assure that having access to a resource will lead to an outcome such as being educated, being able to access information, being able to get a job, becoming financially independent, and so on. This possible outcome is because factors such as societal hierarchies, discriminating practices, norms, and so on may restrict communities' effective opportunities to undertake the actions and activities that they want to engage in. Thus, socio-political and cultural context are powerful influences that we need to understood well (Marcus and Harper 2014).

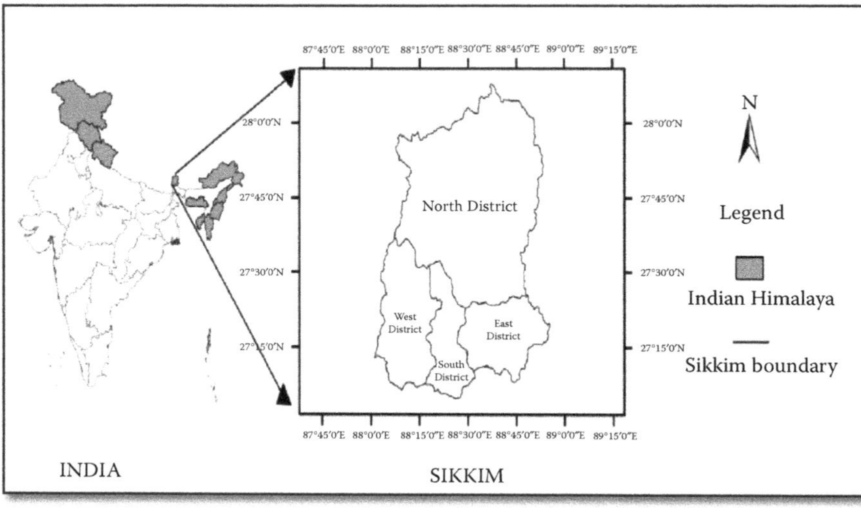

FIGURE 7.1
Map of Sikkim depicting its location in India and as a part of the Eastern Himalayan Region. (From Tambe et al. (2012).)

With this background, this chapter aims to assess the status of education in the East district of Sikkim located in the Eastern Himalayan region of India (Figure 7.1). We intend to provide an understanding of the factors that shape achievements in the education sector in the context of enhancing the adaptation of rural mountain communities of the region. We divided the chapter into four sections: (1) the first section is the introduction, (2) the second section describes the study area selected in the Eastern Himalayan region of India and the methodological approach adopted for assessing the status of education for enhancing the adaptive capacity, (3) the third section presents a discussion on the key findings of the assessment of education in the context of adaptive capacity of selected rural mountain communities of Sikkim, and (4) the fourth section concludes the chapter with our main messages.

7.2 Study Area and Methodological Approach

7.2.1 Background of the Study Area

Sikkim is in the northeastern region of India in the Eastern Himalaya region of India. It is geographically a small state constituting merely 0.22 percent of the total geographical area of India (Government of India 2011). Sikkim shares international boundaries with Tibet, Bhutan, and Nepal. Sikkim is vulnerable to frequent natural disasters such as earthquakes, droughts, and landslides (Government of Sikkim 2009).

Cultivable land is highly scarce in the state as only 11 percent is available for cultivation (Government of Sikkim 2014). Farmers have limited landholdings classified as either as small (1–2 ha) or marginal (below 1 ha). They practice subsistence farming which is rain-fed and is climate-sensitive. The agricultural productivity in Sikkim has remained stagnant due to high dependence on rainfall, limited land holding size, and lack of mechanization along with the frequent occurrence of natural disasters (Government of Sikkim 2009). The economy of the region is largely dependent on agriculture; the non-diversified nature of the economy is also a cause of poverty (Allan 2011).

Low educational achievement in Sikkim is a concern in terms of aggravation of poverty and vulnerability of the rural mountain communities. Although the literacy rate (ability to read and write) is high in the state, that is, 86.55 percent among males and 75.61 percent among females, there is a strikingly high rate of school dropouts (Government of India 2011). Statistics reveal that it is as high as 85 percent, which links to a high rate of unemployment (50 percent) among the youth of Sikkim (National Institute of Rural Development [NIRD] 2012). More than 75 percent of these unemployed youth had secondary school or lower education level (Government of Sikkim 2014). Lack of quality education, economic compulsion, and social issues such as early marriage are the contributing factors towards high school dropout rates in the state (Government of Sikkim 2014). Further, early marriage constrains the possibility of higher education. The average age at which females get married is 21.5 years old, and around 16 percent of the females get married before they turn 18 years old (Government of Sikkim 2014). This early marriage aggravates gender inequality in education and in turn livelihood opportunities. Gender inequality persists in the state as it is a patriarchal society and men have the privilege of ownership of resources, representation in decision-making processes, and availing the benefits of government interventions. In contrast, women are engaged in laborious agricultural activities, and social-cultural norms limit their participation in decision-making processes, for example, their exclusion from the traditional system of governance dzumsa[2] (Government of Sikkim 2014; Subba 2014).

[2] The dzumsa is the traditional administrative system in Sikkim (Government of Sikkim, 2012).

7.2.2 Selection of Study Sites

Among the four districts of Sikkim, East district was selected for the research because the district (1) receives maximum mean annual rainfall (Gangtok, 3494 mm) (National Institute of Disaster Management [NIDM] 2017) and is highly vulnerable to the impacts of climate change with frequent occurrence of landslides during rainy season; (2) is the most populous (46 percent of the state population, that is, 281,293 persons) and has highest population density (297 persons per square kilometer), which is an important parameter of societal vulnerability and represents pressure on resources (Ye et al. 2017); (3) is representative of the state with a literacy rate of 83.85 percent and 50 percent unemployment in rural areas—rural unemployment is an important determinant of lack of adaptive capacity (Tschakert 2007; Government of Sikkim 2014); (4) is vulnerable to impacts on climate-sensitive agriculture sector as agriculture activities of the state are concentrated in the East district because it is in the lower mountain ranges (NIDM 2017) and it has 45 percent of the total agricultural workers (main and marginal) of the state; and (5) has high social diversity as it has the highest percentage of scheduled tribes (38 percent of 206,360), 62.74 percent of the people practice Hinduism, and 25.55 percent follow Buddhism. We selected the district in consultation with the experts of Rural Management and Development Department (RMDD) of the Government of Sikkim at the state and district level.

Among the eight development blocks of the East District, we selected Rhenock and Regoh because they experience medium, medium to high, and high vulnerability to natural disasters. We selected these two blocks in consultation with the block development officers of the development blocks of the East District. We selected Rhenock Tarpin and Sudunglakha gram panchayat units (GPU) from the Rhenock development block, similarly, Dolepche, and Rolep Lamaten GPUs from the Regoh development block in consultation with the field facilitators of the RMDD, Government of Sikkim. The biophysical and socio-economic profile of the four GPUs is in Table 7.1.

7.2.3 Data Collection Methods and Techniques

We used Focus Group Discussions (FGDs) and household surveys as the research methods. For carrying out 16 FGDs in the 4 GPUs (that is, 4 FGDs in each GPU), we selected several representatives from the GPUs. For each FGD, we identified 10–12 representatives of the social groups in GPUs who were in the age group of 25–60 years and had been residing in the GPUs since childhood. We planned FGDs with all-male groups, all-female groups, and mixed groups comprising both male and female representatives (Tables 7.2 and 7.3).

TABLE 7.1

Profile of the Selected GPUs of East District

		East District of Sikkim			
		Rhenock Development Block		Regoh Development Block	
Sl. No.	Indicators	Rhenock Tarpin GPU	Sudunglakha GPU	Dolepchen GPU	Rolep Lamaten GPU
1	Elevation (meters)	1,100	900	1,500	2,000
2	Annual mean temperature (°C)	20	18	17.5	17
3	Annual rainfall (millimeters)	2,900	3,100	3,100	3,200
4	Total households	1,202	322	515	224
5	Total population	6,319	1,720	2,685	1,284
6	Average percentage of Below Poverty Line (BPL) HH	18	38	28	61
7	Average land holding size	0.13	0.29	0.22	1.07
8	Percentage dependent on agriculture	25.08	57.00	43.40	51.50
9	Percentage of illiterate	19	19	17	20
10	Educational achievements (primary level percentage)	30	36	31	40
11	Educational achievements (secondary level percentage)	23	20	27	16
12	Educational achievements (tertiary level: graduate percentage)	9	4	7	3
13	Educational achievements (tertiary level: post graduate school percentage)	0	0	0	0
14	Educational achievements (others percentage)	20	21	19	21

We drew a sample of 35 households from each of the 4 GPUs through stratified random sampling forming a total sample size of 140 households for the household survey. We took a sample size of greater or equal to 30 because it is a statistically large enough population to draw a meaningful conclusion (Levin and Rubin 2007; LaMorte 2016).

TABLE 7.2

Gender-Based Composition of the Participants of Focus Group Discussions (FGDs)

		State: Sikkim				
		District: East Sikkim				
		Rhenock Development Block		Regoh Development Block		
Sl. No.	Gender-Based Composition of Participants of FGD	Rhenock Tarpin GPU	Sudunglakha GPU	Dolepchen GPU	Rolep Lamaten GPU	Total FGDs
1	All male	1	1	1	1	4
2	All female	1	1	1	1	4
3	Mixed (male and female)	2	2	2	2	8
	Total FGDs	4	4	4	4	16

TABLE 7.3

Classification of the Social Group Based on the Caste

		State: Sikkim									
		District: East Sikkim									
		Rhenock Development Block				Regoh Development Block					
		Rhenock Tarpin GPU		Sudunglakha GPU		Dolepchen GPU		Rolep Lamaten GPU		Total No. of Respondents	
Sl. No.	Caste[a]	No.	%	No.	%	No.	%	No.	%	No.	%
1	General	15	42	12	34	6	17	10	29	**43**	**31**
2	Other Backward Caste (OBC)	10	29	15	42	20	57	20	57	**65**	**46**
3	Schedule Tribe (ST)	10	29	8	24	9	26	5	14	**32**	**23**
	Total no. of respondents	**35**	**100**	**35**	**100**	**35**	**100**	**35**	**100**	**140**	**100**

[a] None of the households who interviewed for the research belonged to Scheduled Caste (SC) category because the percentage of SC population were much lower in the East District of Sikkim (4.6 percent) and there were few household belonging to the SC category in the selected study locations (Government of India 2011b).

7.3 Data Analysis

7.3.1 Multidimensional Poverty Assessment Tool

We based the interview schedules which we used to carry out the household survey on the Multidimensional Poverty Assessment Tool (MPAT) framework which the International Fund for Agricultural Development (IFAD) developed (IFAD 2014). It is a suitable analytical framework developed on the understanding that income, or economic growth, is not a sufficient measure of poverty and that poverty is multifaceted, complex, and difficult to define and measure (Streeten and Burki 1978; Streeten et al. 1981; Sen 1999; Bourguignon and Chakravarty 2003; Sullivan 2006; Bossert et al. 2009; Narayan et al. 2009; Alkire and Santos 2010; Alkire and Foster 2011). The significance of using the framework for assessing the status of education is that it considers education as an important component of multidimensional poverty and in turn vulnerability of the individual and of the communities. The MPAT considers analytical framework education as one of the six fundamental needs for the overall well-being of an individual or a community in the context of exposure to shocks posed by natural disasters (Scoones 1998; Department for International Development [DFID] 1999; Cohen 2010; Cohen and Sullivan 2010). Hence, we selected the methodology of MPAT to develop the thematic index of education to understand the various factors that shape educational attainments. We selected three aspects (sub-components) of education: quality, availability, and access (Table 7.4).

We analyzed the responses to the survey questions on the sub-components of education using the standardized procedure of MPAT manual (International Fund for Agricultural Development [IFAD] 2014). The first step of processing the responses to the survey questions on indicator (Table 7.4) was to carry out valuations of each of the responses on a scale of 1–10 to obtain a scaled score x_{ijk} in Equation 7.1, where 1 reflects high poverty and 10 reflects no poverty. In the second step, we aggregated this scaled score of the response to survey questions (indicators) of each sub-component and the weights attached to the respective survey questions w_{ik} (in percentage) by estimating the weighted arithmetic average (Table 7.4 and Equation 7.1). In this way, we obtained a value for each sub-component and each household on a scale of 10–100 (at a higher resolution) (IFAD 2014)

$$y_{jk} = \sum_{i=1}^{L} w_{ik} x_{ijk} \tag{7.1}$$

TABLE 7.4

The Themes with the Sub-Components, the Survey Questions, and Relative Weights Assigned to the Sub-Components

MPAT Components	Sl. No.	MPAT Sub-Components	Indicators of Sub-Components	Relative Importance (Weights) of Indicators of Sub-Components (%)[a]	Relative Importance (Weights) of Cub-Components (%)[b]
Education	1.1	Quality	Teacher pupil ratio in a class	100	31
	1.2	Availability	Education level attained in school and college (school: primary and below, middle, secondary, higher secondary; college: bachelors and above)	100	33
	1.3	Access	Distance to the educational institute	100	36

Source: International Fund for Agricultural Development (IFAD), *The Multidimensional Poverty Assessment Tool: User's Guide*, IFAD, Rome, Italy, 2014.

[a] Weights used for estimating arithmetic weighted average.

[b] Weights used for calculating the geometric weighted average.

where:

y_{jk} is the arithmetically weighted average score for household j in sub-component k

w_{ik} is the weight attached to survey question i in the sub-component k

x_{ijk} is the scaled score for household j in question i in sub-component k (Saisana 2009)

$\sum_i w_{ik} = 1; 0 \leq w_{ik} \leq 1$ (weights represented in percentages in Table 7.4)

The next step was to aggregate the sub-components of education through a geometric weighted average using Equation 7.2. We prefer the geometric weighted average in the final step rather than the arithmetic weighted average because it pulls the index score to the lower value which needs emphasis (Saisana and Saltelli 2010). This result is important because we intend the research to identify the (1) sub-component of education (quality, availability, and access) and (2) GPUs reflecting lower education and in turn lower adaptive capacity. We need to prioritize the sub-component and GPU with the low status of education for enhancing the adaptive capacity.

We consider the arithmetic weighted average from the previous step because x_{ijk}, and w_{ik} are the weights attached to each sub-component (last column of Table 7.4) (Saisana and Saltelli 2010).

$$y_{jk} = \prod_{i=1}^{L} x_{ijk}^{w_{ik}} \tag{7.2}$$

where:

y_{jk} is the geometrically weighted average score for household j of component k

w_{ik} is the weight attached to sub-component I of component k

x_{ijk} is the arithmetically weighted average score for household j in sub-component I of component k (Saisana 2009)

$\sum_i w_{ik} = 1; 0 \leq w_{ik} \leq 1$ (the weights represented in percentages in Table 7.4)

In this way, we developed the thematic index of education at the household level and subsequently at the GPU level. The final step was to develop the thematic index at the district level (y_t). We calculated it by a population-weighted aggregation technique based on population of the 4 GPUs (Table 7.1) as weights (Equation 7.3)

$$y_t = \frac{\sum_{i=1}^{n} w_i x_{it}}{\sum_{i=1}^{n} w_i} \tag{7.3}$$

where:

y_t is the population weighted average score of component (or theme) t for East District

w_i is the population of GPU i as weight

x_{ij} is the thematic index score of GPU i for component (or theme) t

$\sum_i w_{ik} = 1; 0 \leq w_i \leq 1$

The thematic index reflects the level of deprivation, ranging from extreme (0–30 percent), high (30–60 percent), moderate (60–80 percent), to no deprivation (80–100 percent).

7.4 Results and Discussion

We organized the discussion into two sub-sections; the first sub-section comprises the result of the thematic index on the education of the East District of Sikkim. The second sub-section discusses the responses of the communities on educational achievements across different castes, tribes, and gender. This is an important social science research technique to understand the well-being of the communities.

7.4.1 Status of Education in East District, Sikkim

The thematic index value of education for East District of Sikkim is 44 percent, which reflects a high level of deprivation in education.[3] The three sub-components of education, quality (38 percent), availability (47 percent), and access (48 percent) also reflect a high level of deprivation in education as they lie in the range of 30–60 percent (Table 7.5). Among the three sub-components, quality of education has the least score on education index; hence, we need to prioritize it. We measured quality of education through the teacher-pupil ratio in schools: for East District the ratio is 1:60 for senior secondary and 1:30 for primary school. The communities expressed concern over a lack of adequate attention on improving the quality of education in the government initiatives on education such as Sarva Shiksha Abhiyan (SSA) and Right to Education Act (RTE). The focus of these interventions is on providing useful and relevant elementary education for all, and universal enrollment in schools for students between the ages of 6 and 14 years.

Among the sub-components, access to education is relatively better than the quality and availability. However, communities attribute high school dropout rates to poor access. For example, the distance to the nearest primary school is 1–2 km and to secondary school is 3–4 km, while the college and university are 60 km away in Gangtok. The communities expressed that

TABLE 7.5

MPAT Results for Sub-Components Education

Sl. No.	MPAT Components	MPAT Subcomponents	MPAT Subcomponent Score (%)	Interventions (Schemes and Programs)	Focus of These Interventions
1	**Education (44%)** *(fundamental needs)*	Quality Availability Access	38 47 48	Sarva Shiksha Abhiyan (MHRD 2017) Right to education act (Government of India 2009)	To provide useful and relevant elementary education for all Universal enrolment of students between the ages of 6–14 years in schools

[3] The thematic index reflects the level of deprivation: extreme (0–30 percent), high (30–60 percent), moderate (60–80 percent), and no deprivation (80–100 percent) (Cohen, 2010).

most of the children dropped out after class 10 because of long distance and high fees with very little opportunity for provision of an education loan. This situation has resulted in low educational qualification among the communities. For example, only 24 percent of the household respondents were graduates and they cited poor quality of schooling, lack of affordability in terms of income, and lack of accessibility to higher educational facilities as the underlying factors that limited their educational achievements. Poor quality of education and low educational qualification has been a challenging problem for the district as it has led to an increased rate of unemployment among the youths (Narrative 7.1).

The communities expressed their interest in participating in decisions on the future interventions in the education sector to draw attention towards improving the quality of education in the district for better employment opportunities. Their interest in participation is also to make the process of planning and implementation of interventions transparent.

Among the four GPUs, Dolepchen has the highest level of deprivation in education (34 percent), while Rolep Lamaten has least (49 percent) (Figure 7.2). This is also evident in the highest illiteracy rate in Dolepchen (26 percent) (Table 7.1). Dolepchen has extreme deprivation in terms of quality of education (21 percent). Poor quality of education is because of high teacher-pupil ratio (1:40) in the primary schools. Dolepchen GPU also has extreme deprivation of access to educational facilities (27 percent) as the value lies between 0 percent and 30 percent, while there is high deprivation in availability of education (49 percent) (Figure 7.2). Therefore, there is a need for improvement in all the three sub-components of education in Dolepchen.

Sudunglakha GPU has extreme deprivation in the availability of education (29 percent) and has second highest deprivation in education (43 percent). This situation is mainly due to the lack of availability of facilities for higher education because there are only primary schools in the GPU. The respondents perceive private schools to be better than public schools in terms of education facilities and expressed that the private schools are largely affiliated with a Christian missionary. The main reason for not being able to send their children to private school is high fees (Rupees 250–450 per month) which makes it unaffordable due to limited income. The difficulty

NARRATIVE 7.1

"Our children are unable to go for higher education as there are no facilities for higher education in our region, and sending them to Gangtok will only be possible if we have proper income."

—Perceptions on education from FGDs (translated from Nepali and Hindi)

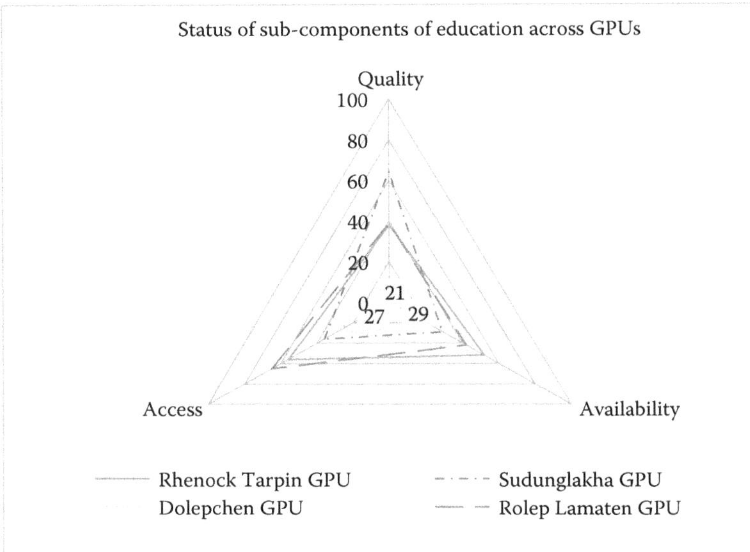

FIGURE 7.2
Status of sub-components of education across GPUs.

in attaining higher education, that is, senior secondary school and college is mainly the long distance (approximately 15 km). Rhenock Tarpin (47 percent) and Rolep Lamaten (49 percent) GPUs have a relatively lower level of deprivation; among the three sub-components, quality of education reflects the highest deprivation in both GPUs, that is, 38 percent and 39 percent, respectively. The respondents from the two GPUs revealed that they have access to primary, secondary, and tertiary (college) education; however, high school dropouts rates are high due to the long distance which has resulted in only a few attaining higher education.

The previous discussions indicate that for overall human development and enhancing their adaptive capacity there is an urgent need to improve the status of education in the East District of Sikkim considering these underlying causes of low achievements in education.

7.4.2 Caste Differentiation in Education

The heterogeneity of the society influences educational achievements which arises due to its social institutions and norms (Unni 2009). According to Sen (1999), learning that stops after providing only basic reading and writing skills is insufficient to advance overall development. Therefore, to look at the actual achievements in education, the perceptions of the communities were captured on issues related to school dropouts and its underlying causes. Although the respondents expressed that there is no discrimination in terms

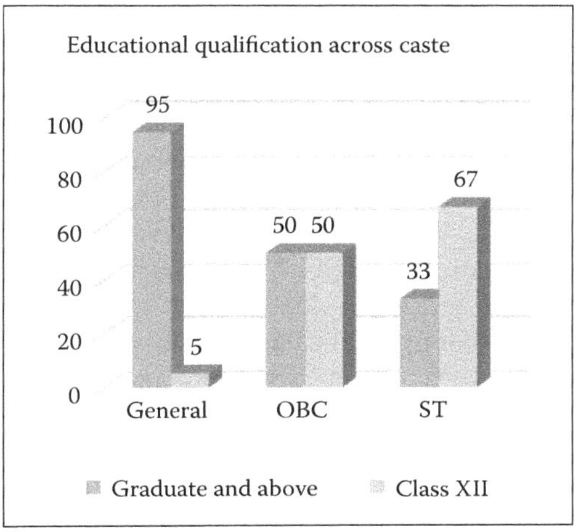

FIGURE 7.3
Perception on educational qualifications likely to attain.

of access to educational facilities, household data shows variation (across caste and tribe) in the conversion of these facilities into actual achievements.

Among the social groups classified based on caste, general caste is more likely to convert the facilities of education into actual achievements. For example, 95 percent of the respondents from the general caste perceived that their children are likely to graduate while the corresponding statistics for the other backward caste (OBC) is only 50 percent and for the scheduled tribes (ST) is 33 percent (Figure 7.3). Further, 67 percent of ST and 50 percent of OBC respondents expressed that their children would generally drop out after high school (class XII or below) leading to lack of achievements from the facilities of education. It is interesting to observe that although the contextual conditions are the same (same distance to the schools and colleges, infrastructure, number of teachers, locations, and so on) across caste and tribe, the capability to convert the facilities of education into actual or desired achievement varies. Studies attribute this variation to various factors, the most prominent factors cited are traditions, history, and culture of the communities. Specific circumstances of individuals or communities shape aspiration and achievements, and intergenerational transfer of knowledge and human resources influences the capabilities of the present generation (Robeyns 2013). To substantiate this, children of upper caste household who have never suffered discrimination in the society would have better access to education and knowledge than compared to those children of Dalit and ST households who have suffered historical deprivation and isolation in the society (Thorat et al. 2016). Caste system among Nepalese society in Sikkim is very similar to that prevalent in a Hindu society where the upper caste Brahmins still maintain social distinction and follow norms of caste purity and impurity (Choudhury 2006). Occupations traditionally depend upon caste hierarchy in the society (Anderson 2011). For example, traditionally in India it is the upper caste who are the large farmers, while OBCs, scheduled castes (SC), and ST are mostly small and marginal farmers, artisans, and laborers (Béteille 1996; Sedwal and Kamat 2008). In the context of Sikkim, although agriculture is the primary activity of the communities (Government of Sikkim 2011), culturally they have their own occupation based on the caste or tribe to which they belong. For example, it is the Brahmins who perform the religious ceremonies while lower caste (OBC, SC, ST) perform low profile jobs such as laborers, metal workers, tailors, or leather making (Choudhury 2006). The literature also suggests that general caste communities are more educated as compared to the lower caste communities such as OBC, ST, and SC communities, who are socially and educationally backward due to a history of deprivation and isolation (Unni 2009; Singh 2016; Thorat et al. 2016). It is important to take into consideration these histories of achievement in parent's education as, according to Lee and Bowen (2006), parent's education is associated with child's achievement in school. For example, "parents who have gone beyond a high school education are found to be more involved with their infants and children than those who did not finish high school...many less educated parents simply have more unmanaged stress in their lives, and this stress interferes with ability and opportunity to interact with their child." (Sclafani 2004, 88). Such households (literate) have higher aspiration for their children due to awareness (such as reading newspaper, listening to news, and so on) on higher returns of investments in education (Unterhalter et al. 2007; Unni 2009). The findings of this research on rural mountain communities of East Sikkim are in agreement with the body of literature on influence of societal structure on education. Figures 7.3 and 7.4 show that the literacy of parents has a positive impact on the education of children. High literacy

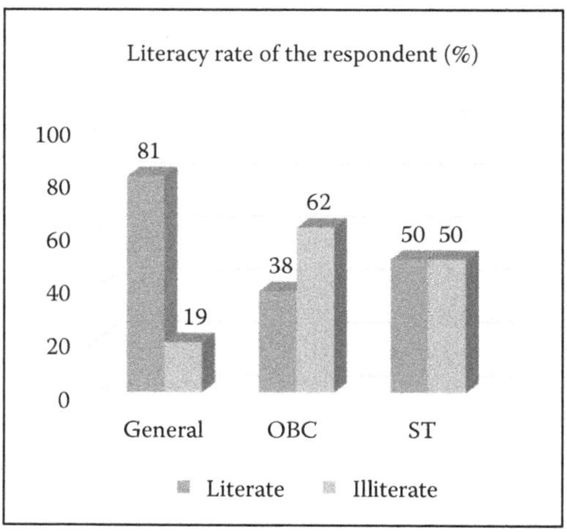

FIGURE 7.4
Perception on literacy rate of the respondent.

among parents from the general caste group has resulted in higher aspiration for their children's education, while low literacy among the parents of OBC and ST groups has resulted in lower aspirations. For example, 62 percent of OBC and 50 percent of ST respondents who are parents were illiterate while only 19 percent of respondents from general caste were illiterate (Figure 7.4).

The low aspiration for education among the parents of OBC and ST household has resulted in higher engagement of their children in agriculture (Figures 7.5 and 7.6). For instance, 83 percent of ST, 68 percent of OBC, and

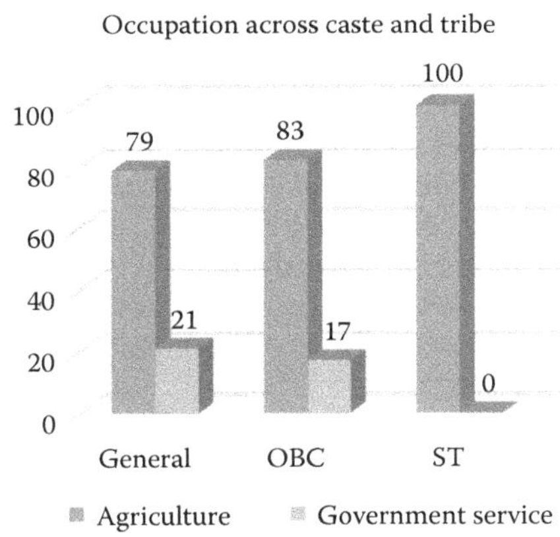

FIGURE 7.5
Occupation across caste and tribe.

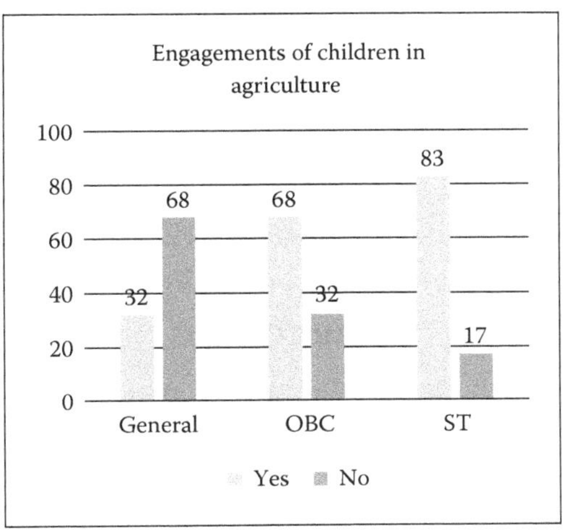

FIGURE 7.6
Engagements of children in agriculture.

only 32 percent of general caste respondents expressed that they engaged their children in agriculture. The findings agree with Laxmaiah and Kachoora (2016) who suggested that engagement of children in agricultural activities leads to high rate of school dropouts resulting in low educational achievements. The additional burden of farming among the children of OBC and ST household has limited their achievements in education. These underlying linkages are likely to be overlooked if capabilities of the individual and communities are not assessed and considered in the process of decision making on education policies. Therefore, the research argues that the interventions should move beyond the provision of resources to improving capabilities of the current and the future generations. After programs convert the resources (e.g., good educational policies, schools, well-trained teachers, and so on) into means, it is also important to have a basket of opportunities and choices for actual achievements. Having larger opportunities improve the overall well-being of an individual and a community (Karimi et al. 2016). For example, someone who can choose between multiple careers is better off than someone limited to one career only, even if both individuals prefer the same career (Karimi et al. 2016). According to Sen (1999), lack of opportunities and choices limits the utilization of the means (for example, education) available to the communities.

The respondent perceived that they lack opportunities and choices to diversify their livelihood due to limited capabilities (Narrative 7.2). They are engaged in agriculture as a compulsion rather than out of choice. A lower potential of livelihood diversification, particularly for people belonging to ST and OBC category, due to lower educational achievement limits their adaptive capacity as livelihood diversification is crucial for enhancing adaptive capacity. Considering this linkage in policy decisions on education would be crucial to overcoming the persisting inequalities in education among the social groups and to enhancing their adaptive capacity.

7.4.3 Gender Differentiation in Education

Socio-cultural norms often restrict the educational achievements of women. The respondents expressed that females have equal access to education and there is no gender differentiation; however, the analysis reveals

NARRATIVE 7.2

"My two sons are graduate, however, they are still engaged in agriculture. They have applied for many jobs but failed to get any. There is no point of going to schools as there are lack of opportunities available in the region"

—Perceptions on education from FGDs (translated from Nepali and Hindi)

striking gender differences in actual achievements. Females have lower educational attainments than males across social categories (Figure 7.7). Further, 83 percent of ST and 77 percent of OBC respondents expressed that women are more likely to drop out after completion of class XII (Figure 7.8). The cultural norms on gender restrict the capacity of women of socially deprived categories (ST and OBC) to convert the facilities of schooling into actual achievements (Unni 2009).

Reports across social categories indicate that the cultural norms of early marriages for women lead to a high rate of school dropouts and a limited utilization of educational resources into actual achievements. The gender

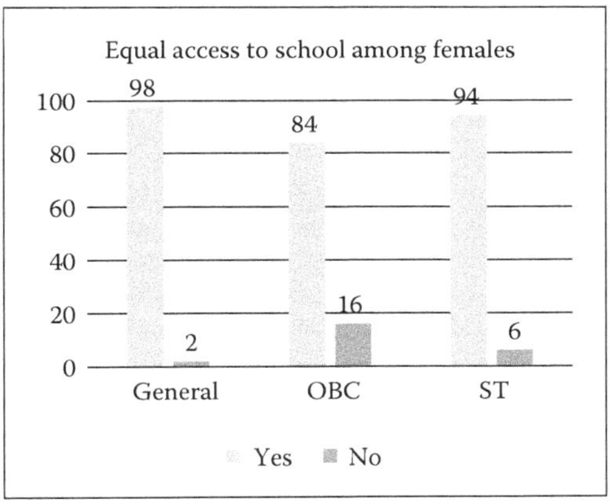

FIGURE 7.7
Equal access to school among females as compared to males.

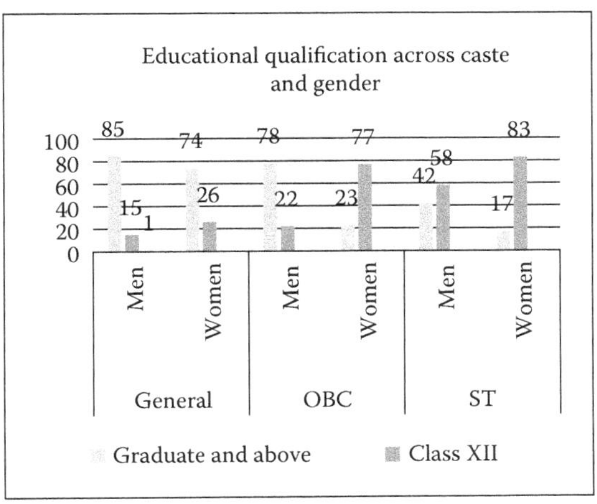

FIGURE 7.8
Perception on educational qualification across caste and gender.

differences also exist in the preference of private schools for a male child because they provide quality education and are relatively expensive and the preference of public schools for a female child (Figure 7.9). Further, a majority of the respondents from the three social categories (91 percent of general, 75 percent of OBC, and 89 percent of ST) expressed that they don't want their daughters to be engaged in an occupation other than agriculture (Figure 7.10). The respondent across social categories shared that there are fewer employment opportunities for females apart from agriculture as compared to males. For example, in the employment generation schemes such

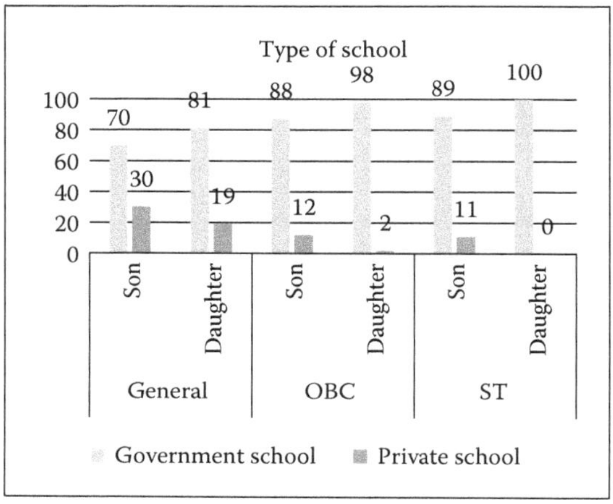

FIGURE 7.9
Type of schooling preferred.

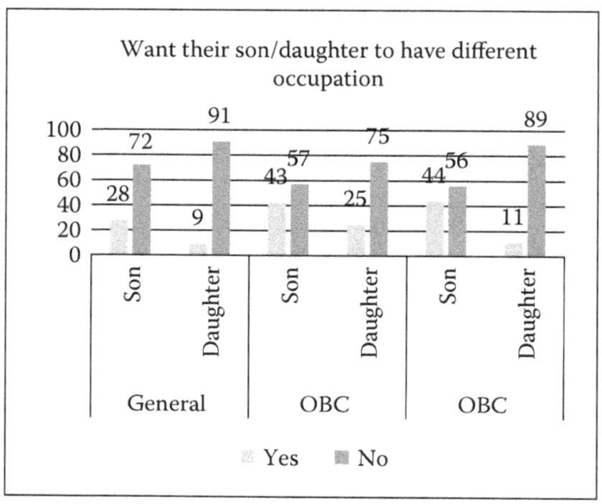

FIGURE 7.10
Different occupation preferred for son and for daughter.

as full form MGNREGA,[4] males hold the job cards while only a few females hold job cards. The females with a job card also failed to find work because of household chores responsibilities. This linkage is important to understand the lower aspiration for the education of women that restricts the conversion of the education into actual achievement.

Women's lack of opportunity for decision-making powers limits their capabilities and results in lack of public opinion and lack of awareness about female-specific concerns and needs in the implementation of interventions in education, health, and livelihoods. This situation is evident because the men of the household make most of the household decisions (65 percent of general, 70 percent of OBC, and 89 percent of ST) (Figure 7.11). Secondly, lack of decision-making power for women has resulted in their lower awareness of the budgets, schemes, and policies implemented in the villages in comparison to males (Narrative 7.3).

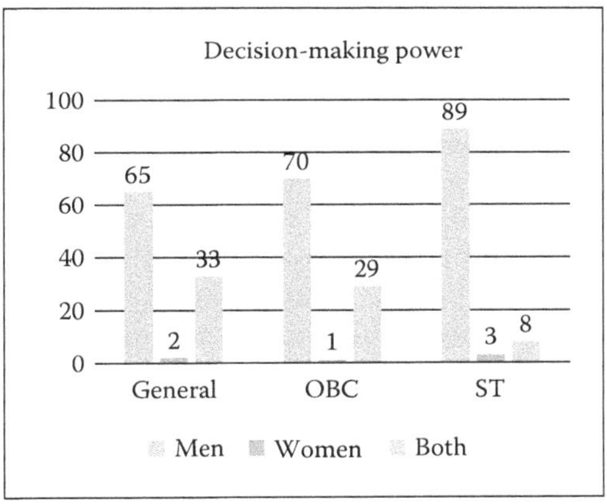

FIGURE 7.11
Household decision taken.

NARRATIVE 7.3

"Although I have been to school and was good at studies, after marriage, the work burden increased after first child birth and other household chores as a result of which I have to discontinue my studies"

—Perceptions on education from FGDs (translated from Nepali and Hindi)

[4] National Rural Employment Guarantee Act, 2005.

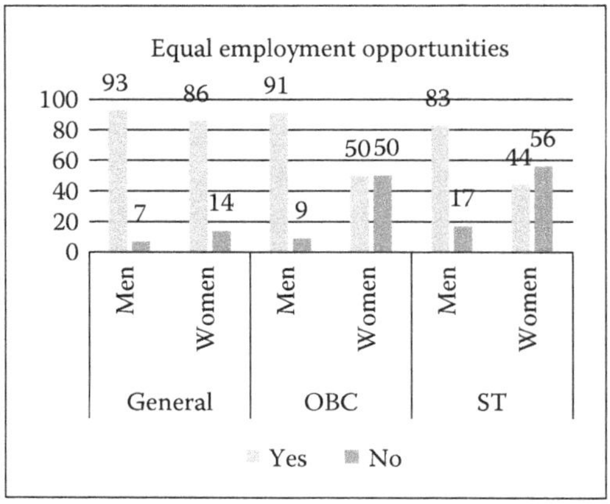

FIGURE 7.12
Equal employment opportunities.

Female respondents have clearly expressed that they do not have equal social and economic opportunities with males (14 percent of general, 50 percent of OBC, and 56 percent of ST) (Figure 7.12). This lack of capabilities among women acts as a deterrent in converting the available resources into actual achievements (Sen 1999) and further leads to discrimination when they enter the labor market (Das et al. 2015). In the context of climate change, educational achievements among women are important because it is the women who are more vulnerable to the impacts of climate change (Denton 2002; Paavola 2008) and deprivation in education further increases the vulnerability of the women of the household (IPCC 2007).

7.5 Conclusion

This chapter aimed to examine the status of education in East Sikkim to understand the factors that shape achievements in education. The thematic index on education developed for East Sikkim revealed that the district faces a high level of deprivation in education, and although attention is needed for improving quality, availability, and access to education, the highest deprivation is in terms of quality of education. This situation hinders livelihood diversification and further lowers adaptive capacity of rural mountain communities. Further, lower educational attainments among the communities have limited their awareness and participation in the interventions on education which prevents them from deriving benefits from the interventions and also influences the decisions on education to enhance adaptive capacity to withstand the impact of natural disasters.

The educational achievements are differentiated by social classification and gender as a result of traditional socio-cultural norms. The assessment revealed that OBC, and ST, and women have lesser adaptive capacity due to limited achievements in education. This situation results in a high dropout rate, a lesser possibility of attaining secondary and tertiary education, and low potential for livelihood diversification. In addition, women's lack of participation in decision-making process due to the socio-cultural norms further marginalizes them, limits their capabilities, and underrepresents their concerns in designing and implementing interventions in education, health, and livelihoods. Therefore, the research emphasizes the importance of designing long-term interventions rather than ad-hoc interventions in education because social change is an essential and time-consuming process for building adaptive capacity. Secondly, it is crucial to integrate the heterogeneity of the society in the design of the intervention because it mediates the conversion of resources (for example, investment in education) into actual achievements (Robeyns 2013). Lastly, the research suggests that interventions should aim beyond the provision of resources towards building capabilities of the rural mountain communities to enhance their adaptive capacity.

Acknowledgments

The Swedish International Development Agencies (SIDA) supported this research; the authors gratefully acknowledge their support. We also acknowledge the Ministry of Tribal Affairs, Government of India for proving organizational support in conducting the study. We express gratitude to the officials of the Rural Management and Development Department (RMDD) of Sikkim for community mobilization and support in making logistic arrangements.

References

Adger, W., N. Brooks, G. Bentham, M. Agnew, and S. Eriksen. 2004. *New Indicators of Vulnerability and Adaptive Capacity.* Norwich, UK: Tyndall Centre for Climate Change Research.

Adger, W. N., and P. M. Kelly. 1999. Social vulnerability to climate change and the architecture of entitlements. *Mitigation and Adaptation Strategies for Global Change* 4(3–4):253–266.

Agrawal, A., and N. Perrin. 2008. *Climate Adaptation, Local Institutions, and Rural Livelihoods.* Ann Arbor, MI: International Forestry Resources and Institutions Program and University of Michigan.

Alkire, S., and J. Foster. 2011. Understandings and misunderstandings of multidimensional poverty measurement. *Journal of Economic Inequality* 9:289–314.

Alkire, S., and M. E. Santos. 2010. *Acute Multidimensional Poverty: A New Index for Developing Countries.* Oxford, UK: Oxford Poverty and Human, Development Initiative Working Paper 38.

Allan, T. 2011. *Virtual Water Tackling the Threat to Our Planet's Most Precious Resource.* New York: I. B. Tauris.

Anderson, S. 2011. Caste as an impediment to trade. *American Economic Journal: Applied Economics* 3:239–263.

Barua, A., S. Katyaini, B. Mili, and P. Gooch. 2014. Climate change and poverty: Building resilience of rural mountain communities in South Sikkim, Eastern Himalaya, India. *Regional Environment Change* 14:267–280.

Béteille, A. 1996. *Caste, Class and Power: Changing Patterns of Stratification in a Tanjore Village.* Berkeley, CA: University of California Press.

Bossert, W., S. R. Chakravarty, and C. D'Ambrosio. 2009. *Multidimensional Poverty and Material Deprivation.* Montreal, Canada: Université de Montréal, Département de sciences économiques.

Bourguignon, F., and S. R. Chakravarty. 2003. The measurement of multidimensional poverty. *Journal of Economic Inequality* 1:25–49.

Brink, E. 2010. *Disasters, Formal Education and Adaptive Capacity.* Lund, Sweden: Lund University.

Choudhury, M. 2006. *Sikkim: Geographical Perspectives.* New Delhi, India: Mittal Publications.

Cohen, A. 2010. The multidimensional poverty assessment tool: A new framework for measuring rural poverty. *Development in Practice* 20(7):887–897.

Cohen, A., and C. A. Sullivan. 2010. Water and poverty in rural China: Developing an instrument to assess the multiple dimensions of water and poverty. *Ecological Economics* 69:999–1009.

Das, S., S. Jain-Chandra, K. Kochhar, and N. Kumar. 2015. *Women Workers in India: Why So Few Among So Many?* Washington, DC: International Monetary Fund.

Denton, F. 2002. Climate change vulnerability, impacts, and adaptation: Why does gender matter? *Gender & Development* 10(2):10–20.

Department for International Development (DFID). 1999. *Sustainable Livelihoods Guidance Sheets.* London, UK: Department for International Development.

Dixit, A., M. Heather, G. Javier, and D. Margaret. 2012. *Ready or Not: Assessing Institutional Aspects of National Capacity for Climate Change Adaptation.* Washington, DC: World Resources Institute.

Gentle, P., and T. Maraseni. 2012. Climate change, poverty and livelihoods: Adaptation practices by rural mountain communities in Nepal. *Environment Science Policy* 21:24–34.

Government of India. 2011. *Census of India.* New Delhi, India: National government publication.

Government of Sikkim. 2009. *State Industrial Profile of Sikkim.* Gangtok, Sikkim: Micro, Small & Medium Enterprises Development Institute.

Government of Sikkim. 2011. *District Census Handbook.* Gangtok, Sikkim: Micro, Small & Medium Enterprises Development Institute.

Government of Sikkim. 2014. *Sikkim Human Development Report 2014: Expanding Opportunities, Promoting Sustainability.* New Delhi, India: Routledge.

International Fund for Agricultural Development (IFAD). 2014. *The Multidimensional Poverty Assessment Tool: User's Guide*. Rome, Italy: IFAD.

Intergovernmental Panel on Climate Change (IPCC). 2001. *Climate Change 2001: The Scientific Basis. Contribution of Working Group I to the Third Assessment Report of the Intergovernmental Panel on Climate Change*. New York: C.U. Press.

IPCC. 2007. *Climate Change 2007: Synthesis Report. Contribution of Working Groups I, II and III to the Fourth Assessment Report of the Intergovernmental Panel on Climate Change*. Geneva, Switzerland: IPCC, p. 104.

IPCC. 2014. Climate change 2014: Impacts, adaptation, and vulnerability. In C.B. Field, V.R. Barros, D.J. Dokken et al. (Eds.), *Part A: Global and Sectoral Aspects. Contribution of Working Group II to the Fifth Assessment Report of the Intergovernmental Panel on Climate Change*. New York: Cambridge University Press.

Jones, L., E. Ludi, and S. Levine. 2010. *Towards a Characterisation of Adaptive Capacity: A Framework for Analysing Adaptive Capacity at Local Level*. London, UK: Overseas Development Institute.

Karimi, M., J. Brazier, and H. Basarir. 2016. The capability approach: A critical review of its application in health economics. *Value in Health* 19(6):795–799.

Lamorte, W. W. 2016. *Central Limit Theorem* [online]. Boston University School of Public Health. http://sphweb.bumc.bu.edu/otlt/mph-modules/bs/bs704_probability/BS704_Probability12.html.

Land, V., and D. Hummel. 2013. Vulnerability and the role of education in environmentally induced migration in mali and senegal. *Ecology and Society* 18(4):14.

Laxmaiah, B., and G. S. Kachoora. 2016. An investigative study on effectiveness of programs reducing high school dropouts in South India: A case of Adilabad District of Telangana. *Elk Asia Pacific Journal of Social Sciences* 2(3):13.

Lee, J.-S., and N. K. Bowen. 2006. Parent involvement, cultural capital, and the achievement gap among elementary school children. *American Educational Research Journal* 43(2):193–218.

Levin, R., and D. Rubin. 2007. *Sampling and Distributions*. 7th ed. New Delhi, India: Prentice Hall.

Magnan, A. 2010. *For a Better Understanding of Adaptive Capacity to Climate Change: A Research Framework*. Paris, France: IDDRI.

Marcus, R., and C. Harper. 2014. *Gender Justice and Social Norms—Processes of Change for Adolescent Girls*. London, UK: Overseas Development Institute.

Mcdowell, G., J. D. Ford, B. Lehner, L. Berrang-Ford, and A. Sherpa. 2013. Climate-related hydrological change and human vulnerability in remote mountain regions: A case study from Khumbu, Nepal. *Regional Environmental Change* 13(2):299–310.

MHRD. 2017. Sarva Shiksha Abhiyan. Department of School Education and Literacy, Accessed January 27, 2018. http://www.mhrd.gov.in/sarva-shiksha-abhiyan.

Muttarak, R., and W. Lutz. 2014. Is education a key to reducing vulnerabilities to natural disasters and hence unavaoidable climate change? *Ecology and Society* 19(1):42.

Narayan, D., L. Pritchett, and S. Kapoor. 2009. *Moving Out of Poverty: Success from the Bottom Up*. Washington, DC: World Bank.

National Institute of Disaster Management (NIDM). 2018. *Sikkim* [online]. National Disaster Risk Reduction Portal. http://nidm.gov.in/PDF/DP/SIKKIM.PDF.

National Institute of Rural Development (NIRD). 2012. *Rural Development Statistics 2011–12*. Hyderabad, India: National Institute of Rural Development.

Norris, F. H., S. P. Stevens, B. Pfefferbaum, K. F. Wyche, and R. L. Pfefferbaum. 2008. Community resilience as a metaphor, theory, set of capacities and strategy for disaster readiness. *The American Journal of Community Psychology* 41(1–2):127–150.

O'Brien, K., and R. Leichenko. 2000. Double exposure: Assessing the impacts of climate change within the context of economic globalization. *Global Environmental Change* 10(3):221–232.

Paavola, J. 2008. Livelihoods, vulnerability and adaptation to climatechange in Morogoro, Tanzania. *Environmental Science and Policy* 11(7):642–654.

Pandey, R., and D. K. Bardsley. 2015. Social-ecological vulnerability to climate change in the Nepali Himalaya. *Applied Geography* 64:74–86.

Robeyns, I. 2013. The capability approach. In F. D. Agostino and G. Gaus (Eds.), *Routledge Companion to Social and Political Philosophy*, pp. 456–466. New York: Routledge.

Saisana, M., and A. Saltelli. 2010. *The Multidimensional Poverty Assessment Tool: Robustness Issues and Critical Assessment*. Luxembourg, UK: Publications Office.

Sapkota, P., R. J. Keenan, J.-A. Paschen, and H. R. Ojha. 2016. Social production of vulnerability to climate change in the rural middle hills of Nepal. *Journal of Rural Studies* 48(Supplement C):53–64.

Sclafani, J. D. 2004. *The Educated Parent: Recent Trends in Raising Children*. Hartford, CT: Praeger Publishers.

Scoones, I. 1998. *Sustainable Rural Livelihoods: A Framework for Analysis*. Brighton, UK: Institute of Development Studies.

Sedwal, M., and S. Kamat. 2008. *Education and Social Equity with a Special Focus on Scheduled Castes and Scheduled Tribes in Elementary Education*. Brighton, UK: Consortium for Research on Educational Access, Transitions and Equity.

Sen, A. 1999. *Development as Freedom*. New York: Oxford University Press.

Sharma, E., N. Chettri, K. Tse-Ring, A. B. Shrestha, F. Jing, P. Mool, and M. Eriksson. 2009. *Climate Change Impacts and Vulnerability in the Eastern Himalayas*. Kathmandu, Nepal: ICIMOD.

Shrestha, A. B., C. P. Wake, P. A. Mayewski, and J. E. Dibb. 1999. Maximum temperature trends in the Himalaya and its vicinity: An analysis based on temperature records from Nepal for the period 1971–1994. *Journal of Climate* 12(9):2775–2786.

Singh, M. 2016. Reservation amidst the Din of 'Development'. *Economic and Political Weekly* 51(38):65–69.

Singh, R. K., R. C. Shrivastava, and T. K. Mukherjee. 2009. Community-based sustainable natural resources management and development in Northeast India. *Current Science* 10:19–21.

Smith, J. B., R. J. T. Klein, and S. Huq. 2003. *Climate Change, Adaptive Capacity and Development*. London, UK: Imperial College.

Streeten, P., and J. S. Burki. 1978. Basic needs: Some issues. *World Development* 6(3):411–421.

Streeten, P., J. S. Burki, U. M. Haq, N. Hicks, and F. Stewart. 1981. *First Things First: Meeting Basic Human Needs in Developing Countries*. New York: Oxford University Press.

Striessing, E., W. Lutz, and A. G. Patt. 2013. Effects of educational attainment of climate risk vulnerability. *Ecology and Society* 18(1):16.

Subba, B. 2014. Women "Quest" for empowerment in Sikkim's society. *International Journal of Scientific and Research Publications* 4(9):1–5.

Sullivan, C. 2006. Do investments and policy interventions reach the poorest of the poor? In P. Rogers (Ed.), *Water Crisis: Myth or Reality?* London, UK: Taylor & Francis Group.

Tambe, S., G. Kharel, M. L. Arrawatia, H. Kulkarni, K. Mahamuni, and A. K. Ganeriwala. 2012. Reviving dying springs: Climate change adaptation experiments from the Sikkim Himalaya. *Mountain Research and Development* 32(1):62–72.

Thorat, S., N. Tagade, and A. K. Niak. 2016. Prejudice against reservation policies. *Economic and Political Weekly* 51(8): 61–69.

Tschakert, P. 2007. Views from the vulnerable: Understanding climatic and other stressors in the Sahel. *Global Environmental Change* 17(3):381–396.

United Nations Framework Convention on Climate Change (UNFCCC). 2007. *Impacts, Vulnerabilities and Adaptation in Developing Countries.* p. 68, New York: UNFCCC.

United Nations International Children's Fund (UNICEF). 2003. *UNICEF Annual Report*. New York: UNICEF.

Unni, J. 2009. Gender differentials in education: Exploring the capabilities approach. *Economic and Political Weekly* 44(9):111–117.

Unterhalter, E. 2009. Education. In S. Deneulin and L. Shahani (Eds.), *An Introduction to the Human Development and Capability Approach.* London, UK: Earthscan.

Unterhalter, E., R. Vaughan, and M. Walker. 2007. *The Capability Approach and Education.* Nottingham, UK: The University of Nottingham.

Wamsler, C., E. Brink, and O. Rantala. 2012. Climate change, adaptation, and formal education: The role of schooling for increasing societies' adaptive capacities in El Salvador and Brazil. *Ecology and Society* 17(2):19. doi: 10.5751/ES-04645-170202.

Waston, C. 2014. *Understanding Changing Social Norms and Practices around Girls Education and Marriage.* London, UK: Overseas Development Institute.

Ye, Y., X. Wei, X. Fang, and Y. Li. 2017. Social vulnerability assessment by mapping population density and pressure on cropland in shandong province in China during the 17th–20th century. *Sustainability, MDPI* 9(7):1171.

8

Governing National Actions for Global Climate Change Stabilization: Examples from India[1]

Joyashree Roy, Shyamasree Dasgupta, Duke Ghosh, Nandini Das, Debalina Chakravarty, Debrupa Chakraborty, and Sudipta De

CONTENTS

8.1 Introduction 138
8.2 Analytical Framework 139
8.3 Alignment of National and Sub-National Governance in Green Initiatives in India 141
8.3.1 Green Initiatives in the Industry Sector: Case of Perform, Achieve, and Trade 141
8.3.1.1 Description of the Example Background 141
8.3.1.2 Articulated Roles of Different Levels of Governance 143
8.3.1.3 Analysis of Actual Roles Played by Different Levels of Governance 143
8.3.2 Potential Green Initiative in the Power Generation Sector: Case of Poly-Generation 145
8.3.2.1 Description of the Example and Background 145
8.3.2.2 Analysis of Articulated Roles of Different Level of Governance 146
8.3.3 Green Initiative in the Transport Sector: Case of Solar e-Rickshaws 147
8.3.3.1 Description of the Example and Background 147
8.3.3.2 Analysis of Articulated Roles of Different Level of Governance 148

[1] We presented an earlier version of this chapter as part of the Indian-European Multi-level Climate Governance Research Network (IECGN) at "Beyond 2015: Exploring the Future of Global Climate Governance," in De Rode Hoed, Amsterdam. Organizers: Institute for Environmental Studies, VU University, Amsterdam. Authors acknowledge with thanks comments from network partners on various occasions.

8.3.4 Green Initiatives in the Building Sector: Case of Standards and Labeling for Home Appliances 151
8.3.4.1 Description of the Example and Background 151
8.3.4.2 Articulated Roles of Different Levels of Governance 151
8.3.4.3 Analysis of Actual Roles Played by Different Levels of Governance 152
8.3.5 Adaptation Strategies and Multilevel Governance in India 153
8.3.5.1 Description of the Example and Background 153
8.3.5.2 Analysis of Actual Roles Played at Different Level of Governance 154
8.4 Concluding Remarks 155
References 157

8.1 Introduction

The worldwide scientific consensus is that there is need for extensive mitigation effort to reduce the concentration of greenhouse gases in the atmosphere to avoid dangerous anthropogenic interference with the climate system (United Nations Framework Convention on Climate Change [UNFCCC] 2015; Intergovernmental Panel on Climate Change [IPCC] 2014a). Climate is a global good and hence managing climate change calls for identification of global goals. Many countries have ratified limiting the average temperature increase to 1.5°C or below 2°C above the pre-industrial level. Over the past couple of decades nations and groups of nations have used scientific information to create awareness about the impending crisis, to chart out policy innovations, and to deliberate on the need for radical transformative changes. In 2015, the bottom-up commitments reached at the Paris Agreement by ratifying countries concerning emissions mitigation through the implementation of Nationally Determined Contributions (NDCs) marked a major milestone achievement in the process of national alignment to a global goal. This agreement was follow up to the similar agenda put forward during the Cancun Agreement in 2010. In the lead up to Paris Agreement, the commitment that the government of India had announced included a reduction in the emissions intensity of its gross domestic product (GDP) by 33 to 35 percent below 2005 levels and an increased share of non-fossil fuel to 40 percent with respect to power generation. India plans to achieve these targets by 2030 GOI 2015; Busby and Shidore 2017). Countries with specific pledges of emission reduction are no doubt a very important global initiative; but scientific reports suggest that even these pledges will remain significantly inadequate to deliver the desired climate goal (IPCC 2014; Sanderson et al. 2016). Not only that, pledge and actions of a country may not be in conformity with each other and hence making the climate goal even more difficult to achieve. The complexity that causes such divergence between pledge and action arises with discussions about "how to achieve such targets?" Emission mitigation does not automatically follow the ratification of international accords and pledges. At the bottom-most level it depends on actions of individual consumers and producers. Between the global goal and individual actors, multiple levels of other governances play important roles in shaping the behavioral response of the bottom-level actors. We consider a more expanded definition of governance to look beyond the governments and refer to the decisions makers at various levels. The complexity of interaction between these various levels of governance is, however, unprecedented and a major determinant of how the process evolves. First, it is the national government (such as India) or the supra-national body (such as the European Union) that states the pledge regarding its contributions to the global mitigation effort. Then the national governments formulate policies to achieve the pledge. While ideally the domains of these policies should define the role of each of the sub-national components of the governance structure at various levels so that bottom-level actors can scale up and integrate their responses in the process of emission mitigation, they are often overlooked. The challenge of climate change mitigation has therefore provided an opportunity for a major social experiment on how national governments can enforce national level policies to achieve a global goal working with of various levels of governance in a fragmented world.

The research question of this chapter is how important is it to align the policies and actions of various levels of governance within the national level to make commitments and initiatives successful? The chapter explores this issue based on selected examples of greenhouse gas mitigation efforts in the energy supply and energy demand sectors (industry, buildings, and transport) and in the context of climate change adaptation in India. Section 8.2 presents the analytical framework of the chapter. In Section 8.3 we analyze one selected case from each of the sectors mentioned and provide concluding remarks in Section 8.4.

8.2 Analytical Framework

We use the concept of vertical integration as defined in the literature of multilevel governance (Corfee-Morlot et al. 2009; Hooghe and Marks 2003) and modify it to accommodate the analysis of the downstream alignment of various levels of governance within the federal structure of India. Keeping in mind the global climate goal and the national pledge, we consider a five-level vertical climate governance structure in India. The structure consists of (1) the central government; (2) state governments; (3) local governments

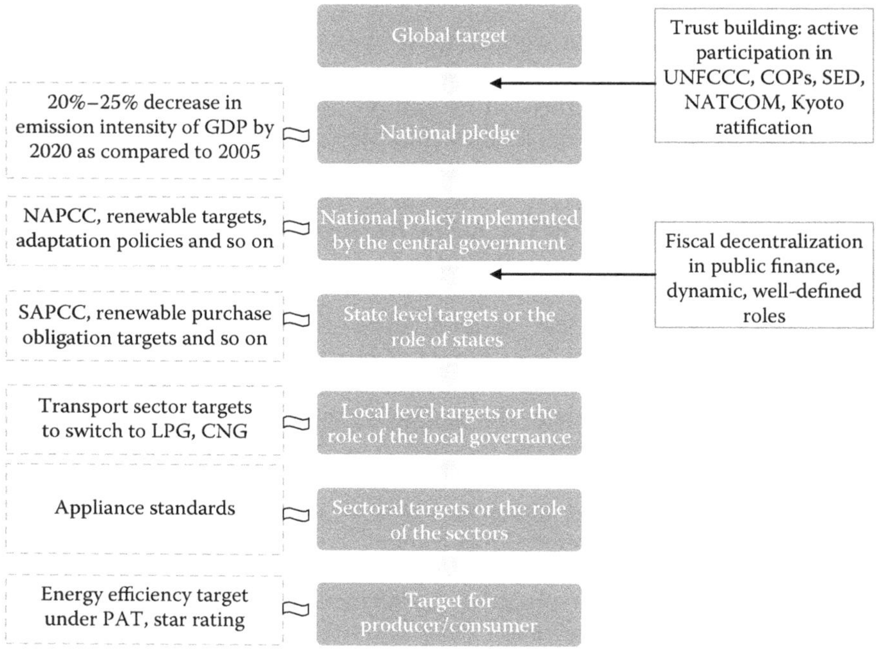

FIGURE 8.1
The vertical alignment of the climate governance structure in India.

(such as panchayats, municipalities, and so on); (4) the sectors (such as power generation, industry, building, and so on); and (5) individual producers and consumers (such as the production unit of an integrated steel plant or multiple small paper production units, and so on) (Figure 8.1).

However, not all levels of governance are relevant, nor do they take part in the implementation of any specific policies and required actions. Sometimes the link between two levels of governances is target based and therefore articulated more objectively and directly between the two levels. Sometimes the roles of governance levels are articulated not in the form of targets but as a supporting mechanism. Depending on the nature or the rigor of the articulated role, we identified them as direct, indirect, or insignificant. The action performed by each level may not, however, match the articulated role. We followed a two-stage analysis for the selected climate policies or actions. The first stage analyses the specific policy or climate action along with the nature and quality of articulation of roles by various governance levels. The second stage analyses how the actor at specific governance has actually performed at the articulated role level. In Sections 8.1 and 8.5 we included narratives of the climate policy or action and articulation of roles. However, the second stage analysis showing how the actors performed the roles are missing in Sections 8.2 and 8.3 because they were only emerging options and were at a stage too premature for analysis. But our analysis highlights a priori the assessment of articulated roles which can strengthen future implementation. In our two-stage analysis, we, therefore, evaluate the nature or rigor of the articulated role of a specific level of governance and the actual role performed by the same. This analysis helps to develop a unique framework for critical analysis of the missing or inappropriate articulation of roles and/or actions at various levels of governance for specific examples. The analysis clarifies why incumbent governance structures may resist new, climate-compatible rules, regulations, and technologies (Raven 2012; Rehman et al. 2010). By citing the examples, we also show why multilevel governance is a strategy that is crucial for speeding up transition through scaling up of low-carbon green experiments.

8.3 Alignment of National and Sub-National Governance in Green Initiatives in India

Connecting national pledges with the global goals is easier as compared to downstreaming the pledge to action by variety of actors. The process depends on the existing institutional environment, its flexibility and barriers, strategic prioritization, and so on. In India, several examples show that despite strong functioning of fiscal federalism in financial matters, vertical alignment in climate governance is emerging at a varying pace across sectors (GOI 2014). Here we deliberately choose examples from different sectors so that they cover successful, less successful, and new experiments with high climate and non-climate benefit potentials to justify the need for conscious efforts towards multilevel climate governance.

8.3.1 Green Initiatives in the Industry Sector: Case of Perform, Achieve, and Trade

8.3.1.1 Description of the Example Background

Perform, achieve, and trade (PAT) is an energy efficiency policy under the National Action Plan on Climate Change (NAPCC) (2008) implemented in eight energy intensive industrial sectors in India: (1) Aluminum, (2) Cement, (3) Chlor-Alkal, (4) Fertilizer, (5) Iron and Steel, (6) Pulp and Paper, (7) Textile, and (8) Power. The genesis of PAT started with the Energy Conservation Act (ECA) (2001) (GOI 2001; Dasgupta and Roy 2015). Under the provisions of the ECA, the government required the bigger energy consuming units, for example, the Designated Consumers (DC), to have their energy uses audited by accredited energy auditors, implement techno-economically viable recommendations made by the energy managers, and comply with energy norms. Later, under PAT, speeding up transition the government identified 478 DCs in eight industries as the most energy intensive units and placed them under the purview of mandatory reduction in Specific Energy Consumption (SEC) during 2012–2013 and 2014–2015. After the compliance period, the Bureau of Energy Efficiency (BEE) received a report of over achievement and under-achievement of the DCs. The government awarded the Energy Saving Certificates (ESCerts) to DCs that overachieved. The government gave the underachievers two options, either purchase ESCerts from the overachievers or pay a penalty (GOI 2012). The PAT could accelerate energy efficiency improvement for facilities accounting for 25 percent of India's fossil fuel use and reduce 6.68 million tons of oil equivalent (MTOE) resulting in reduction in CO_2 emissions by 25 million ton per year by the end of the first compliance period compared to a business-as-usual period (GOI 2010; Roy 2010). This calculation is approximately 1.4 percent of India's projected total annual CO_2 emissions in 2015 (IEA 2009; Roy 2010) (Figure 8.2).

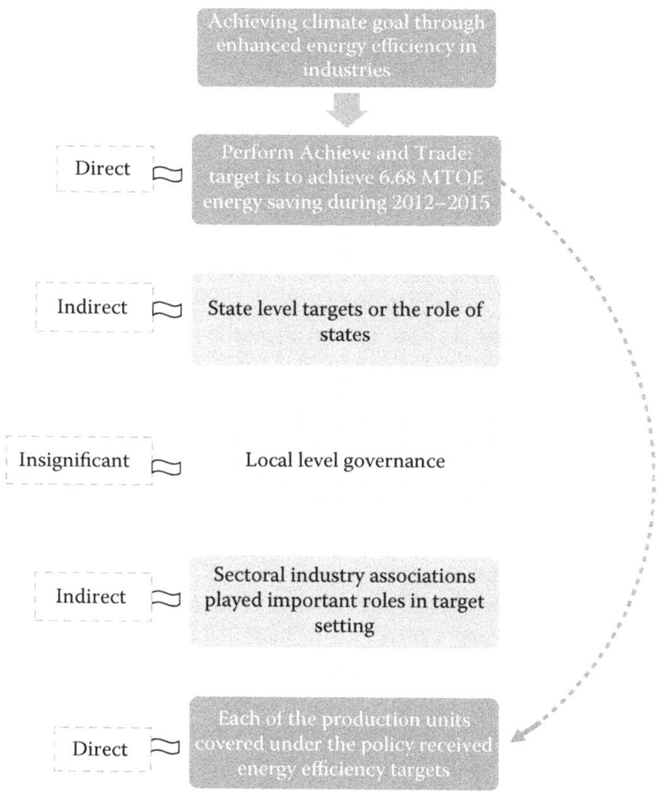

FIGURE 8.2
Role of different levels of governance in implementation of perform, achieve, and trade (PAT).

8.3.1.2 Articulated Roles of Different Levels of Governance

In 2001, the ECA provided the legal framework, institutional arrangement, and regulatory mechanism at the central and state levels to initiate the energy efficiency programs. The government created the BEE under the central Ministry of Power to act as the nodal agency for implementation with well-articulated roles. The government entrusted the BEE to benchmark the energy use in the DCs and identify the SEC targets while requiring the DCs to report back to the BEE directly at the end of the compliance period. The ECA also mandated establishment of state-level statutory bodies, State Designated Agency (SDAs), to *"coordinate, regulate, and enforce the provisions"* of the ECA in the state (ECA 2001). The ECA also mandated states to constitute a state Energy Conservation Fund (SECF) with contributions from the central government to utilize in implementation of energy efficiency improvements, promoting renewable energy, and sustaining awareness creation activities (BEE 2012). However, the policy formulation under PAT linked the individual production units, situated in the bottom most level of the governance structure, directly to the central government through BEE (the top-most governance) only with indirect interventions of and collaborations with the state government to facilitate the process. However, the policy did not articulate roles for the local level governance. Although the policy did not establish sector-specific targets, it acknowledged the role of individual industry associations in determination of SEC targets.

8.3.1.3 Analysis of Actual Roles Played by Different Levels of Governance

In practice, BEE played the role of a nodal implementing agency as articulated. SDAs collaborated with the BEE to identify the DCs and to set benchmarks. But the main role of the BEE remained capacity building through organizing workshops and training programs, involving the energy managers and energy auditors and DCs, appraising roles as per the mandate of the ECA, and creating media and awareness campaigns. Although creation of the SEFC was one of the articulated roles, the creation of such funds did not happen. However, during the development phase of PAT, the BEE conducted extensive consultations with the DCs through sectoral and even plant level discussions in various states.

The process of vertical integration of the governance structure in implementation of PAT made two significant contributions. First, although there were no state-specific targets, the participation of state as a facilitator contributed greatly. The capacity building and campaign program organized by the states resulted in a verified reduced capacity of 1,065 MW during the eleventh plan in India (BEE 2012). Second, the 2-year-long consultation between the national and sub-national actors including the state, industry association, and the plant level actors helped the program to emerge significantly. Initially, the objective of the policy was to set up a single process-wise benchmark for each of the sectors. It was through these interactions and the survey conducted by the BEE that the BEE finally reached the decision for plant-specific SEC targets owing to the vast diversity even within a sector. This plan shows that the integration of various levels of governance at the early stage of policy implementation, even in case of a mandatory policy, plays an important role in identifying innovative solutions to address country- and industry-specific realities. The government expects that this program, therefore, will increase the rate of compliance by increasing preparedness, demonstrating flexibility in policy design, and securing participant engagement (Climate and Development Knowledge Network [CDKN] 2013).

There are, however, some missing links. The delay in creating the SECF indicates a failure of the linking mechanism between the central- and the state-level governance which can significantly reduce the pace of energy efficiency drive in the country. Secondly, PAT is a top-down approach with the central governance imposing the targets and the actions happening at the bottom level. As a result, so far, the government does not acknowledge the other green initiatives conducted by industries beyond the purview of energy efficiency. The state cannot realize the scaling up of these green initiatives unless governed strategically through adding bottom-up components in the existing energy efficiency polices. The face-to-face interview with cement, paper and paperboard, and steel plants revealed that industries are interested in multiple ways to enhance climate responsiveness beyond energy efficiency because it enhances their economic competitiveness (Roy et al. 2017). For example, one of the paper plants mentioned that the plant has the largest wind farms in the country and exports power to the state power grid. In many cases, the paper plant sold the sludge it generated in the paper-making process to the local board mill where they convert it into sundry papers. The mini cement plants use the lime sludge and fly ash generated by the paper mills. Some of the paper plants converted fly ash into cement and bricks and sold them in the local markets and for road construction in the nearby areas. Many plants are also burning the sludge and using it as a substitute for coal (both domestic and imported) in the production process. Steel plants are switching fuel (petroleum coke, coal dust, coal gas, and so on), substituting virgin material by enhancing use of waste materials (scrap materials put back into the furnace, re-using rejected pipes and slags in the production process), improving energy conservation (use of heat treatment furnace to reduce the use of consumed energy and use of iron ore fines instead of iron ore lumps), and minimizing wastes by using by-products like Blast Furnace Slag, Blast Furnace dust, Basic Oxygen steel Slag (BOS) dust, Steel Furnace Slag, and so on. To encourage and incentivize these experimental efforts not only is the bottom-up linkage important but also the horizontal linkage between the different industrial sectors, thus the boundary issue is likely to emerge in a significant manner.

8.3.2 Potential Green Initiative in the Power Generation Sector: Case of Poly-Generation

8.3.2.1 Description of the Example and Background

The Indian power sector is presently coal dominated and hence CO_2 intensive. It also depends on centralized power generation in large (mega or ultra-mega) plants and distributed through national grid. However, the need for the introduction of more renewable power is crucial for future energy security and even more for reducing CO_2 emission from power sector (Hiremath et al. 2009; GOI 2015). Decentralized renewable power generation matching local demand with resources is emerging as a prospective solution for low-carbon power in the future (Banerjee 2006). Integrating these small-scale and intermittent generations into a grid maintaining the quality of power is a critical technological challenge and advanced countries are actively exploring this (Assadi et al. 2014). However, for India ample scope also exists for grid-independent, intermittent, and decentralized small-scale power generation specifically for areas with severe shortage of grid power (Bhattacharyya 2006). A Multi-utility system, or poly-generation, matching local resource with demand may be even more efficient (Jana et al. 2017). A country may blend grid-integrated decentralized power with advanced technological options with this grid-independent option with proper planning. An optimization between power from the national grid and from off-grid sources using local resource is important for India. India urgently needs policy planning at the national level in line with the sub-national level up to the smallest unit (that is, municipality or panchayats) for this purpose because then resources will be local but outputs may be used either locally (that is, off-grid) or nationally (that is, grid-integrated) for an optimum solution and to match the global commitment. But a no less important challenge is how the existing institutional structure with the regulatory mechanism, actors/stakeholders, and routines will accommodate and resist the multi-utility system.

Poly-generation is the process of combining multiple utility outputs with one or more input(s) in a single integrated unit (Chicco and Mancarella 2008; Qian et al. 2009; Bose et al. 2013). Poly-generation targets the integration towards higher energy efficiency as well as environment friendliness. Plants can use biomass for the generation of electricity, heat, and motor fuel with a low impact on environment (Joelson and Gustavsson 2010). We need to recognize and proactively implement complex combinations of institutional roles. While the emphasis is more on technology development and transfer issues without regard to governance issues, poly-generation goes well beyond cogeneration: power utility and heating utility, trigeneration (that is, power, heating, and cooling). Poly-generation includes delivery of the services of three or more utilities (that is, electricity, utility heat, drinking water, biofuel, fertilizers, and so on) using one or more resources (fossil or renewable) in a single well-integrated unit.

It is beneficial economically, environmentally, and emerging as a very useful future sustainable solution for development.

8.3.2.2 Analysis of Articulated Roles of Different Level of Governance

Developing poly-generation systems which can transform the rural growth in India needs creative technological and social innovation. The government should base the initial conceptualization of the system on identified needs of utilities supported by the knowledge of available or near-future technologies and local resources. Systematic integration of these technologies into an economically feasible, environmentally benign, and socially acceptable "sustainable" unit requires evolution through optimization. Deciding control logics for the various demands of different utilities from the same unit is the next step. The plant may need repeated iterations involving these concepts to finally configure the system through simulation. Subsequent development may include component design, procuring and manufacturing of these components, and integration of these components in the final installation. The plant should follow it with an inspection of its operation to assess its real performance.

Capacity of poly-generation units is another important technological issue. Depending mostly on available resources for input, the plant can decide the size of the unit. The plant will most likely decide the capacities of output utilities based on local needs. However, one of the utility outputs of poly-generation is most likely to be power. The plant may use this power locally off-grid, or integrate it through the smart grid. The government may decide the policy for the capacity of poly-generation, specifically including power, at the local as well as the national level. However, technological challenges for integrating poly-generation through smart grid needs even more governance strategies.

Implementation of poly-generation in coal-based power generation in any country apart from political priorities involves several issues such as energy policy, technology policy, existing fuel policy, and finally climate policy. Poly-generation requires the government to build an institutional structure. The framework for decentralized poly-generation as partial replacement of coal-based power primarily depends on the initiatives and policy formulation of the central government to promote development of a low-carbon energy system agenda to mitigate the impact of climate change. the administration of power and new and renewable energy can play a key role in this development. In the provincial level, apart from investment-sharing, providing protection to sustain this niche is crucial because this effort is vulnerable to existing market force in the power generation and supply sector. Giving tax benefits can be one such protection to support sustainability. The state government also plays an important role in formulation of institutional framework and its maintenance (Figure 8.3).

As mentioned earlier, local need is an important feature of this new technology operation and local cooperation must provide maintenance of this

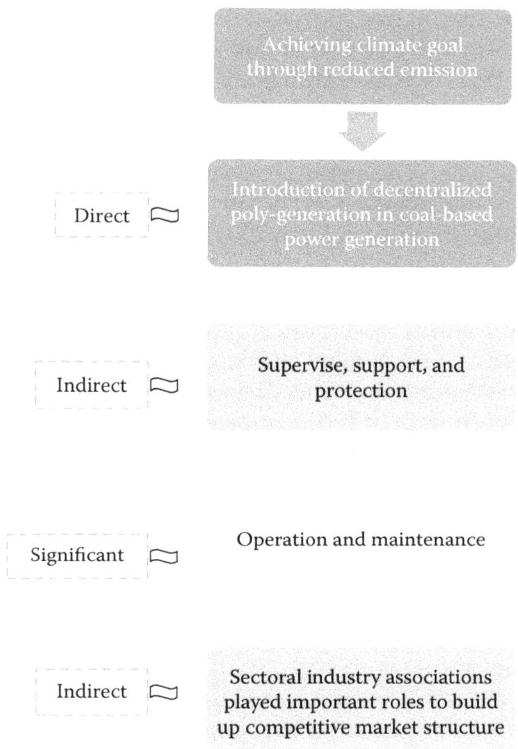

FIGURE 8.3
Role of different levels of governance in implementation of poly-generation.

system. The state government must consider resource mobilization and fuel supply variation according to the availability at ground level.

Poly-generation of energy utility supply has a climate-change mitigation potential with respect to low emission; however, economic feasibility is an equally important aspect. Currently in India, despite the high potential for prospects, poly-generation of multiple utility services products requires encouragement from a designated authority for proper promotion.

8.3.3 Green Initiative in the Transport Sector: Case of Solar e-Rickshaws

8.3.3.1 Description of the Example and Background

The non-motorized, cycle rickshaws (and vans) have been a popular intermediate mode of transport (IMT) in the Indian mobility space. They perform dual services—moving people and goods. For some rural areas and small towns, these vehicles are the major means of transport (Rajvanshi 2002). For cities and towns with a network of narrow lanes, these vehicles are perhaps the most effective means of transport for short distance point-to-point mobility. Further, this IMT provides employment to millions of poor Indians and, more importantly, this IMT is non-emitting and non-polluting. Since 2007–2008, technology research institutions and entrepreneurs in India had taken up a strategic initiative to transform the human power-driven rickshaws into motorized vehicles. The objective was to enhance speed and usability together with reducing stress on the health of the drivers of cycle rickshaws/vans (Maji et al. 2010; Begum and Sen 2004).

During the last decade, India has been the site for least three prominent experiments; the Solar Electric Rickshaw (Soleckshaw) was the most recent experiment. The Central Mechanical and Electrical Research Institute (CMERI), Durgapur, West Bengal, initiated this experiment in 2007–2008 under the CSIR-800 mission. The National Research and Development Corporation, a government of India enterprise, had the authority to transfer the technology to the Indian entrepreneurs on a non-exclusive basis.

Soleckshaw is a pedicab that the driver can drive "by a combination of manual power and electric motor to optimize the utilization of the electric power" (Maji et al. 2010). The payload of the vehicle is 210 kg (considering two passengers and one driver). The average speed of the vehicle is approximately 15 km/h. The combination of motor and manual pedaling optimizes the use of manual power and electric power. The scientists associated with the innovation claim that "as the speed of the vehicle can match the speed of the automobiles [auto rickshaws] in busy streets and narrow lanes, the vehicle has the potential to replace a substantial number of public transport vehicles [auto rickshaws]...thus resulting in the reduction of air pollution and emission" and noise pollution (Maji et al. 2010). While designing the pedicab, the scientists also aimed at making this vehicle a truly zero-emission vehicle. Therefore, they factored the idea of using solar power in the design to charge the batteries. Initially, the hood of the vehicle served as a frame to accommodate the solar panels. In the recent designs, scientists added two charging options: (a) solar charging stations; (b) a main power supply source using a battery swapping mechanism. These two options provided the rickshaw drivers with flexibility in charging the batteries.

Soleckshaw thus aimed for sustainability gains. These gains were not only environmental gains (reducing air and noise pollution) but also economic gains (increased earnings of the poor rickshaw pullers and increasing the welfare of the poor). Therefore, the experiment was also a sustainability experiment (Berkhout et al. 2009).

8.3.3.2 Analysis of Articulated Roles of Different Level of Governance

The innovation by CMERI, a central government-owned institution, was under the CSIR-800 mission. The mission pledges to foster inclusive growth in India by improving the quality of life through the application of science and technology for 800 million citizens of the country who live "at the bottom of the pyramid" (CMERI 2010). Therefore, from the start the innovation was inclusive and pro-poor. Thereafter, the then-Central Minister for Science and Technology and the Chief Minister of Delhi launched Soleckshaw in New Delhi, the capital of India, on October 2, 2008. Incidentally, the date of launch is significant; it marks the birth anniversary of Mahatma Gandhi. Thus, the government further positioned the innovation as an object of national importance and stressed the welfare and environment-related objectives of the innovation. Initially, a fleet of seven vehicles started operating in New Delhi on a pilot basis. Together, in Kolkata, CMERI distributed five vehicles, free of charge, to a local non-governmental organization (NGO) that started operating the vehicles through local rickshaw pullers. These were the ways the central government tried to popularize the technology. However, our research reveals that people abandoned the Soleckshaws operating in Kolkata within a short span of time because the vehicles could not navigate the flooded streets of the city during the rainy season.

In 2010, the New Delhi Municipal Council (NDMC) planned to replace the fleet of manually operated rickshaws through the Electric Rickshaws. The design of the vehicles ensured that the motor power was less than 250 W and so the Soleckshaw was not under the purview of the Central Motor Vehicles Rules of 1989, and did not require licensing for the driver (Maji et al. 2010). This decision by NDMC proliferated the number of these vehicles in Delhi. Meanwhile, the private sector also started manufacturing similar vehicles, and thus similar low-cost variants were available in the market. The low-cost variants did not have solar panels. Thus, there was a marked departure from the original idea of promoting a zero-emission vehicle, and so, because of electrical charging, the extent of indirect emission was considerable. The new design adopted by the private manufacturers also augmented the seating capacity considerable: double or triple when compared to the conventional cycle rickshaw. This feature, along with availability of subsidies and non-requirement of driving license, attracted many poor persons to drive e-rickshaws in the streets of New Delhi. However, as the drivers did not receive training to drive these motorized vehicles, the vehicles became a source of many road accidents in New Delhi (TOI 2014). In April 2014, the Central Ministry for Road Transport imposed a temporary ban on the electric rickshaws. Finally, in July 2014, a Delhi court banned electric rickshaws in New Delhi (Ghosh et al. 2016; WSJ 2014) (Figure 8.4).

The social networks are the venues of expectations and promises, articulators of renewed requirements and demand, sources of resources, and enablers in the respect of dissemination of learning across (and between) actors and locations (Raven 2012). The prevalent regime networks are insufficient as they use the incumbent rules and routines and are more prone to tread the existing trajectory and desist from exploring the new ones (Raven 2012; Rehman et al. 2010).

The overall configuration of the actors involved in the conventional cycle rickshaws is fragmented, informal, and unorganized. However, the network

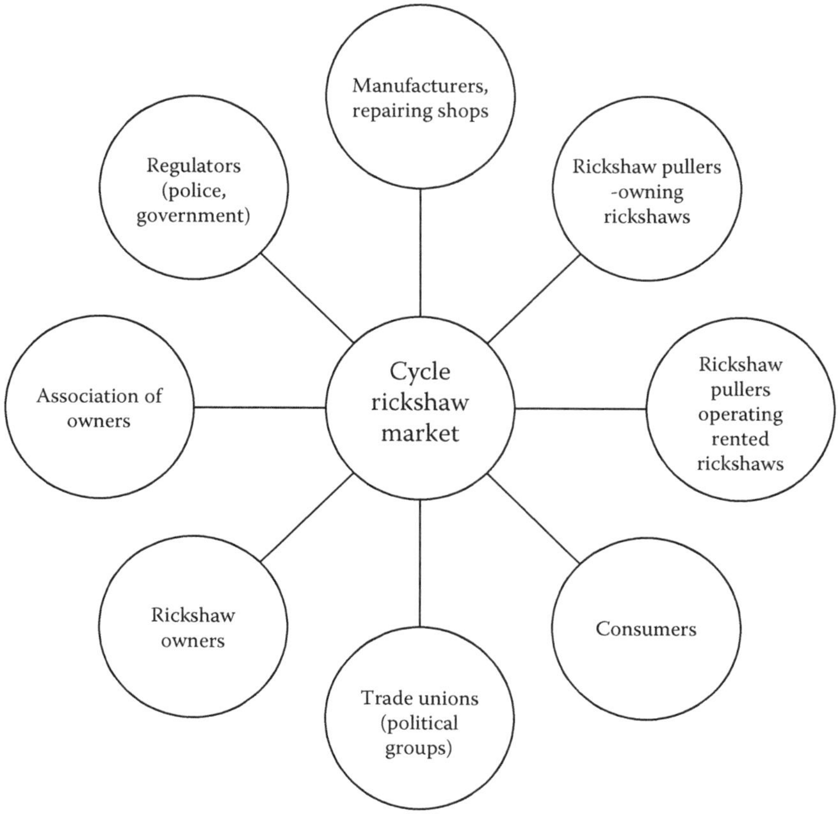

FIGURE 8.4
Actor network of the conventional manual rickshaws.

is complex. Considering governance, the Municipal Authorities govern the incumbent regime of the cycle rickshaw. Collusive oligopoly becomes a source of power in the bargaining process among the actors. Binding agreements between the actors hardly exist and this leads to the demonstration of rent-seeking behavior among the powerful groups within the set-up. The rickshaw pullers hold the weakest position in the whole system.

Given the complexities, technical and legal, in the case of electronic rickshaws, the network needs to be formal and regulated. The network needs to take into cognizance the issue of governance. Because the electric rickshaws has a motor, with adequate incentives to increase the capacity of the motor, it could fall under the auspices of the Central Motor Vehicles Rule. And, with this, the conflict between the decisions and of the local and federal governments emerges. This is what happened in New Delhi (Ghosh et al. 2016).

Equally important is the issue of governing the niche through a more integrated approach that involves technology partners, trade unions, and so on. Then the training for handling these vehicles belongs to the drivers. The erstwhile manual "rickshaw pullers" are not adequately trained to handle these motorized vehicles. Hence, the network needs actors responsible for training and licensing of drivers as per the federal laws of the state. The electric rickshaw niche in India, although showing promises of sustainability gains, suffers a temporary setback because of an incomplete network of actors.

8.3.4 Green Initiatives in the Building Sector: Case of Standards and Labeling for Home Appliances

8.3.4.1 Description of the Example and Background

The Honorable Minister of Power launched a scheme on May 18, 2006 that initially involved 12 types of equipment and appliances, that is, (1) room air conditioners; (2) fluorescent tube lights; (3) frost free refrigerators; (4) distribution transformers; (5) induction motors; (6) direct cool refrigerators; (7) electric storage type geysers (gas-fired water heaters); (8) ceiling fans; (9) color televisions; (10) agricultural pump sets; (11) LPG stoves; and (12) washing machines, of which the first four had mandatory labeling from January 7, 2010 (MoP; GOI 2012). The other appliances are presently under the voluntary labeling phase. The intent of the energy efficiency labeling programs is to reduce the energy consumption of electric home appliances without diminishing the services they provide to consumers (BEE 2012). The government expects that this program will generate energy savings and thereby the cost savings, reduce capital investment in energy supply infrastructure, enhance the product quality, strengthen the competitive markets, build position for domestic industries to compete in such markets where norms for energy efficiency are mandatory, remove indirect barriers to trade, reduce carbon emission, and help meet climate change goals. The star rating indicated the efficiency standard with ranges from 1 to 5 in the increasing order of energy efficiency. Energy star specifications differ with each item and the Environmental Protection Agency (EPA) or the Department of Energy set them (BEE 2012; TERI 2010). For example, the efficiency parameter of standards and labeling (S&L) for air conditioning is the energy efficiency ratio (EER), lumens per watt for fluorescent tube light (FTL), service value as per Indian Standard (IS) 374:1979 for ceiling fan, kilowatt hour (kWh) per kilogram (kg) for a washing machine, and so on. The most recent addition to the list of labeled products is corporate average fuel consumption (CAFC) standards for passenger cars in January 2014.

8.3.4.2 Articulated Roles of Different Levels of Governance

In May 2006, BEE launched an S&L program for electrical home appliances. The bureau initiated the program with technical information support from appliance-producing companies. BEE was supposed to be a nodal implementing agency from policy making to implementation. The program expects the

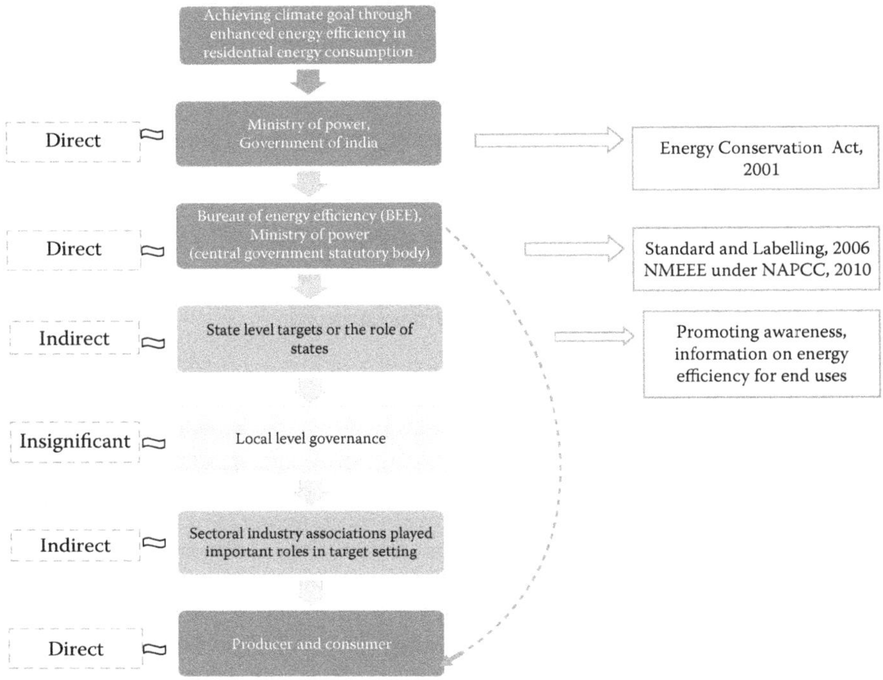

FIGURE 8.5
Role of different levels of governance in implementation of standard and labeling.

SDA of each state to promote energy efficiency at the end user level so that the end user can make an informed choice on appliances and thus the demand for efficient appliances arises in the market. The stakeholders of this program are appliance-manufacturing firms and the end users (Figure 8.5).

8.3.4.3 Analysis of Actual Roles Played by Different Levels of Governance

Information dissemination is slow and limited through SDAs as they did not feel the ownership of the program. On the other hand, BEE was more successful than the SDA initiatives. In 2010, BEE again tried to emphasize the program through then National Mission for Enhanced Energy Efficiency (NMEEE) under National Action Plan on Climate Change (NAPCC). NMEEE aims to strengthen the market for energy efficiency by creating conducive regulatory and policy regime and has envisaged fostering innovative and sustainable business models to the energy efficiency sector. In the Indian market, the manufacturers of mandatory star-rated appliances are gradually accepting the regulation. Some large manufacturers of geysers, color televisions, and washing machines are taking this star-rated policy as a market advantage as being in the forefront of efficiency innovation. When government impose standards, it causes significant additional declines in residential energy demand and an increase in net cumulative benefit (Levine et al. 1995; Hirst and Eto 1995; McKane et al. 2009).[2] Typically, energy efficient devices have higher initial cost or purchase prices, but lower operating costs than their less energy efficient counterparts (Levine et al. 1995; Hirst 1984). This higher initial cost could be the reason for low market demand for efficient appliances and thus for the low production of such appliances. The star rating indicator differs for different appliances. It is not clear to the consumer for some appliances, such as TFL, ceiling fan, and so on, how much electricity unit and therefore electricity bill they can save by switching to 5-star rated appliances from no-star rated or 1-star rated appliances.

8.3.5 Adaptation Strategies and Multilevel Governance in India

8.3.5.1 Description of the Example and Background

India occupies 2.4 percent of the world's geographical area with a large percentage of its land under agriculture and nearly 69 percent falls within dry land (Bhandari et al. 2007). The Intergovernmental Panel on Climate Change fifth assessment report (IPCC 2014b) clearly warns that projected climate change would lead to further water shortage and quality problems. The government of India and several state governments have launched various programs to conserve soil and water resources for food security. The government of India has begun to recognize the benefits of watershed management for a more long-term perspective of combating climate change (Mitra et al. 2008). Successful implementation of the Integrated Watershed Management Program (IWMP) can help in the restoration of ecological balance, increase in productivity levels in rain-fed areas, and improve living conditions in rural areas (Bhandari et al. 2007), (Nagaraja and Ekambaram 2015).

The government of India's Ministry of Rural Development implemented three watershed programs: Integrated Wastelands Development Programme (IWDP), Drought Prone Areas Programme (DPAP), and Desert Development Programme (DDP). The government of India launched the DPAP in 1973–1974 to tackle the special problems faced by those areas that are constantly afflicted by drought conditions, DDP launched in 1977–1978 to mitigate the adverse effects of desertification, and IWDP has been under implementation since 1989–1990 (GOI 2011). In 2008, the government of India combined these three programs under the Integrated Watershed Management Programme (IWMP). Although IWMP is not a direct program under climate change policies, it has a co-benefit on climate resilience building. IWMP is a classic

[2] Many developing country governments view energy conservation as a luxury and have shown little interest (Reddy 1991).

example of federal policy implementation where every stakeholder has their own specified role to perform as per guidelines of the central government. Central and state government share the cost of these programs: 75:25 ratios for DPAP and DDP and for IWDP 11:1. There always exists an overlapping zone between developmental activity and adaptation measures. Initially, the government of India started IWMP as a part of capacity building initiatives to improve conditions in the rain-fed regions, but it has a long-term perspective of adapting to climate change because all activities related to water, soil, and land conservation contribute to adaptation. Almost all the adaptation activities somehow link with the development , and mitigations activities are part of the sustainability goal of that development. Apart from promoting overall economic development, mitigating the adverse effects of extreme climatic conditions and restoring ecological balance by harnessing conserving natural resources is other prime focus of this program (GOI 2001).

8.3.5.2 Analysis of Actual Roles Played at Different Level of Governance

The Indian federal system governs IWMP at multilateral levels. It is basically a top-down approach where policy originates centrally and implements at the grass roots level. The government of India sanctions the projects as per procedure and guidelines and implements programs mainly through the Zilla Parishads (ZPs)/District Rural Development Agencies (DRDA). However, wherever it is expedient in the interest of Watershed Development Programs, any department of the state government or autonomous agencies of central government or state governments with the approval of the Department of Land Resources, Government of India, may implement the projects (GOI 2001). The role of state government is basically monitoring and evaluating the progress of the project (Figure 8.6).

An effective and efficient multilevel governing system needs a well-defined institutional framework. Planning, finance, implementation, monitoring, and coordination are the specified governing aspects in this model and the system vests all these roles upon various institutes of the federal structure at multiple levels. The most significant feature of the governing system of IWMP is that here every sub-national institution has a specified role to play which is neither direct nor indirect. The central government plays the role of a planner and one of the main financing sources. The role of state government is to monitor and to provide part of the financial resources. Zilla Parishad, the third tier of Indian federal structure, has the role of implementation and coordination with the central and state government. IWMP follows the vertical integration in its governing structure. Considering the growing importance of the private sector in Indian context, the scope of improvement of IWMP lies in the integration of the private sector as a new stakeholder in form of public private partnerships (PPPs) (Rahaa et al. 2013). Introduction of the PPP in this framework would contribute in efficiency improvement through inter-sectoral linkage.

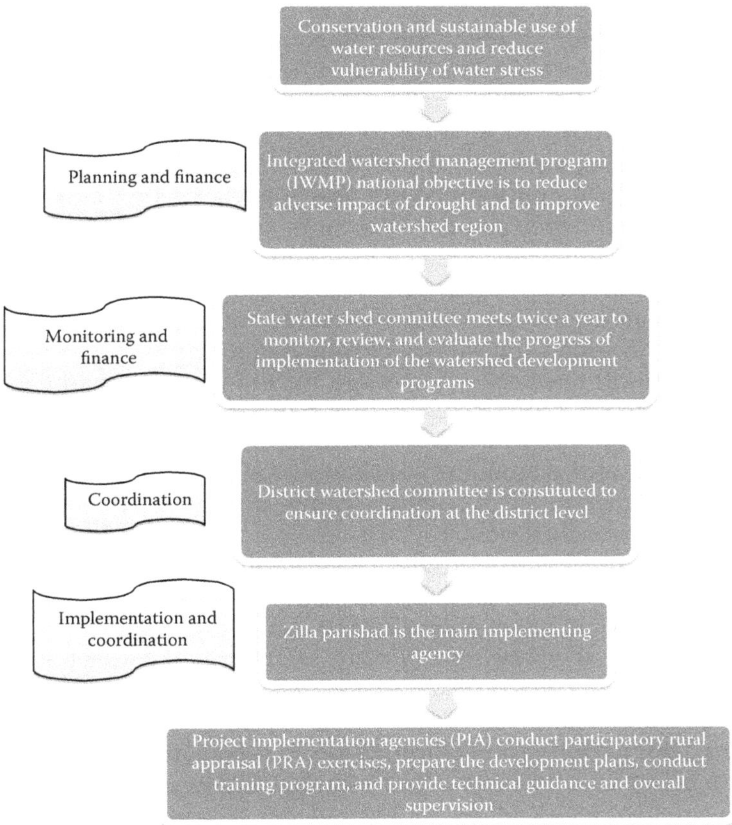

FIGURE 8.6
Role of different levels of governance in implementation of the Integrated Watershed Management Program (IWMP).

8.4 Concluding Remarks

In climate governance both upstream and downstream alignment in multilevel governance structure is crucially important to deliver the goals. Given the Paris Agreement, where national actions will build up the pathway for achievement of the global goal, downstream vertical integration in multilevel climate governance becomes a necessary step and, in some sense, will be sufficient as well. In India, several examples show that despite strong functioning of fiscal federalism for a long time, downstream vertical alignment in climate governance is still very weak. One major reason behind this situation may be the current political positioning of the country through voluntary pledges in international negotiations on climate change.

The other important reason is the lack of appropriate institutions (only exception is BEE) for strategic prioritization and goal-oriented vertical integration aiming for economic growth that is compatible with a serious climate change agenda. The various programs are still at experimental (for example, biodiesel in mobility sector, e-rickshaws) or at discussion (for example, poly-generation) levels. The industry and household appliance sectors planned and worked out their policies better; however, with more strategic action they could realize more untapped technical potential. Target-oriented multi-actor alignment within nation states are assuming greater importance with the planning of new emerging technological solutions to deliver energy security and low-carbon futures. For example, while technical optimization between power from national grid and from off-grid sources using local resource is important for India, policy planning at the national level in line with that at the sub-national level down to the local level-municipality or panchayats is even more crucial to deliver the service. Analysis of past and ongoing efforts shows the need for recognition and proactive implementation of complex combinations of institutional roles with strategic scientific inputs.

Coordination of global, national, sub-national goals, and actor-level actions are indispensable. So, the resolution of how to make political decisions compatible with scientific and economic realities is a major issue. Since national circumstances vary, aligning national actions with a global goal will be a real challenge and the question is can it be a matter of international negotiation? How far a country coordinates nationally determined mitigation actions through sub-national alignment efforts can be an integral part of a negotiation strategy. Emergence of a successful governance structure depends on how global efforts are necessitating a well-defined governance structure. Political positioning of the national governments in a global platform also matters. The role of social and governance innovation is usually less understood or given less priority than technological innovation at the policy-making level. This situation is apparent when we notice technology centric policies; technology transfer debates find more priority than debate, policy, or action for new governance mechanism design or institutional arrangement design and coordination among various line departments within existing administrative structures. So how to make political decisions compatible with scientific, (for example, the case study of poly-generation), economic (competitiveness, fuel switch, waste recycling), social (Soleckshaw), and institutional realities (PAT) is a major research question and problem to be resolved. Studies quite often do not discuss technology and governance together. This is business as usual. Social, economic, and institutional configuration need to accompany any technological and financial configuration since the market mechanism alone cannot govern everything. We need a transformative change through a systemic transformation and not only through technology change or transfer.

References

Assadi, M., M. M. Majoumerd, K. Jana, and S. De. 2014. Intelligent biogas fuelled distributed energy conversion technologies: Overview of a pilot study in Norway. ASME 2014 Gas Turbine India Conference (pp. Paper No. GTINDIA2014-8231, pp. V001T08A005; 10 pages). International Gas Turbine Institute.

Banerjee, R. 2006. Comparison of options for distributed generation in India. *Energy Policy*, 34(1):101–111.

Begum, S., and B. Sen. 2004. Unsustainable livelihoods, health shocks and urban chronic poverty: Rickshaw pullers as a case study. *CPRC Working Paper* 46, pp. 1–26.

Berkhout, F., D. Angel, and A. Wieczorek. 2009. Asian development pathways and sustainable socio-technical regimes. *Technological Forecasting & Social Change* 76:218–228.

Bhandari, P. M., S. Bhadwal, and U. Kelkar. 2007. Examining adaptation and mitigation opportunities in the context of the integrated watershed management programme of the Government of India. Mitigation and Adaptation Strategies for Global Change.

Bhattacharyya, S. C. 2006. Energy access problem of the poor in India: Is rural electrification a remedy? *Energy Policy*, 34(18):3387–3397.

Bose, A., K. Jana, D. Mitra, and S. De. 2015. Co-production of power and urea from coal with CO2 capture: Performance assessment. *Clean Technologies and Environmental Policy* 17(5):1271–1280. doi: 10.1007/s10098-015-0960-7.

Bureau of Energy Efficiency (BEE). 2012. http://beeindia.in/documents/12.pdf. (accessed October 25, 2014).

Busby, J. W., and S. Shidore. 2017. When decarbonization meets development: The sectoral feasibility of greenhouse gas mitigation in India. *Energy Research & Social Science* 23:60–73.

CDKN. (2013). http://r4d.dfid.gov.uk/PDF/Outputs/CDKN/India-PAT_InsideStory.pdf. (accessed October 25, 2014).

Chicco, G., and P. Mancarella. 2008. A unified model for energy and environmental performance assessment of natural gas-fueled poly-generation systems. *Energy Conversion and Management* 2008:2069–2077.

CMERI. 2010. CMERI, Quadrennial Report (2006–2010). Durgapur: Central Mechanical Engineering Research Institute.

Corfee-Morlot, J., L. Kamal-Chaoui, et al. 2009. Cities, climate change and multilevel governance. OECD Environment Working Papers; Paris, p. 120. Accessed 2 December 2009. https://www.oecd.org/governance/regional-policy/44232263.pdf.

Dasgupta, S., and J. Roy. 2015. *Designing PAT as a climate policy in India: Issues learnt from EU-ETS*. Chapter 16 in *Nature, Economy and Society: Understanding the Linkages*, N. Ghosh, P. Mukhopadhyay, A. Shah, and M. Panda (Eds.), 315–328. New Delhi, India: Springer. doi:10.1007/978-81-322-2404-4.

Dyre, M., S. Frazer, and K. Price. 2006. *Confronting climate change and emissions reduction in the energy sector. Emissions trading EU-ETS experience and lessons for New Zealand*. Paper Presented at the 8th Annual New Zealand Energy Summit, New Zealand. http://www.climatechange.co.nz/publications/Emissions%20Trading%20-%20EU%20ETS%20Lessons%20for%20NZ.pdf. (accessed January 17, 2014).

Ghosh, D., F. Sengers, A. J. Wieczorek, B. Ghosh, J. Roy, and R. Raven. 2016. Urban mobility experiments in India and Thailand. In *The Experimental City*, J. Evans, A. Kavonen, and R. Raven (Eds), p. 122. Abingdon, UK: Routledge.

Government of India (GOI). 2001. *The Energy Conservation Act of 2001*. New Delhi: Ministry of Power, Government of India. http://powermin.nic.in/acts_notification/pdf/ecact2001.pdf. (accessed October 24, 2014).

GOI. 2001. *Guidelines for Watershed Development*. New Delhi: Department of Land Resource, GOI-Ministry of Rural Development. http://dolr.nic.in/dolr/guidewd.asp. (accessed January 14, 2014).

GOI. 2010. India: *Taking on Climate Change. Post-Copenhagen Domestic Actions*. New Delhi: Ministry of Environment and Forests, Government of India.

GOI. 2011. *National Watershed Development Programme*. http://www.archive.india.gov.in/sectors/agriculture/index.php?id=7. (accessed January 13, 2015).

GOI. 2012. *PAT: Perform Achieve and Trade*. New Delhi: Ministry of Power, Government of India.

GOI. 2014. *The Fourteenth Finance Commission (FFC)–Implications for Fiscal Federalism in India*. New Delhi: Government of India.

GOI. 2015. *India's Intended Nationally Determined Contribution: Working Towards Climate Justice*. http://www4.unfccc.int/submissions/INDC/Published%20Documents/India/1/INDIA%20INDC%20TO%20UNFCCC.pdf. (accessed March 14, 2017).

Hiremath, R. B., B. Kumar, P. Balachandra, N. H. Ravindranath, and B. N. Raghunandan. 2009. Decentralised renewable energy: Scope, relevance and applications in the Indian context. *Energy for Sustainable Development* 13(1):4–10.

Hirst, E. 1984. Household energy conservation: A review of the federal residential conservation service. *Public Administration Review* 1984:421–430.

Hirst, E., and J. Eto. 1995. *Justification for Electric-Utility Energy-Efficiency Programs*. Oak Ridge National Laboratory.

Hooghe, L., and G. Marks. 2003. Unraveling the central state, but how? Types of multi-level governance. *American Political Science Review* 97(2):233–243.

International Energy Agency (IEA). 2009. *World Energy Outlook 2009*. Paris: OECD/IEA. http://www.worldenergyoutlook.org/media/weowebsite/2009/WEO2009.pdf. (accessed October 20, 2014).

IPCC. 2014a. Climate change 2014: Synthesis report. Contribution of working groups I, II and III to the fifth assessment report of the intergovernmental panel on climate change. Geneva, Switzerland: IPCC.

IPCC. 2014b. *Climate Change 2014: Impacts, Adaptation, Vulnerability*. Cambridge, UK: Cambridge University Press.

Jana K., A. Ray, M. M. Majoumerd, M. Assadi, and S. De. 2017. Poly-generation as a future sustainable energy solution–A comprehensive review. *Applied Energy* 202:88–111.

Joelson, J. M. and L. Gustavsson. 2010. Reduction of CO2 emission and oil dependency with biomass-based polygeneration. *Biomass and Bioenergy*, 34(7):964–987.

Levine, M. D., et al. 1995. Energy efficiency policy and market failures. *Annual Review of Energy and the Environment* 20(1):535–555.

Maji, P. K., P. S. Banerjee, A. J. Banerjee, and S. Maity. 2010. Electric motor-assisted pedal driven tricycle. *International Journal of Electric and Hybrid Vehicles* 2(3):202–210.

McKane, A., I. Rhyne, A. Lekov, L. Thompson, and M. Piette. 2008. Automated demand response: The missing link in the electricity value chain. 2008 ACEEE Summer Study on Energy Efficiency in Buildings. Pacific Grove, CA: American Council for an Energy-Efficient Economy.

Mathur, A. 2011. *Energy Synergy with Dr. Ajay Mathur*. New Delhi: Bureau of Energy Efficiency, Government of India. www.indiaworldenergy.org/ppt/NMEEE.pps. (accessed January 2014).

Mitra, A., S. Chpde, A. Kumar, and S. A. Wajih. 2008. *Climate change adaptation activities in India*. Gorakhpur, India: Gorakhpur Environmental Action Group.

MoP (Ministry of Power), Government of India. 2010. *Energy Efficiency*. Accessed December 2014. http://powermin.nic.in/Energy-Efficiency.

Nagaraja, B., and G. Ekambaram. 2015. A critical appraisal of integrated watershed management programme in India. *IOSR Journal of Humanities and Social Science*, 17–23.

Qian, Y., L. Jingyao, Z. Huanga, A. Kraslawskib, J. Cuia, and Y. Huanga. 2009. Conceptual design and system analysis of a poly-generation system for power and olefin production from natural gas. *Applied Energy* 2009:2088–2095.

Rahaa, D., H. Osbahra, and C. Garfortha. 2013. Does watershed development implemented through public private partnership empower women? A case review from Rajasthan, Western India. *Journal of Cleaner Production* 60:18–33.

Rajvanshi, A., 2002. Electric and improved cycle rickshaw as a sustainable transport system for India. *Current Science* 83(6):703–707.

Raven, R. P. 2012. Analysing emerging sustainable energy niches in Europe: A strategic niche management perspective. In *Governing the Energy Transition*, G. P. Verbong and D. Loorbach (Eds.), 125–151. New York: Routledge.

Reddy, A. K. N. 1991. Barriers to improvements in energy efficiency. *Energy Policy* 19(10):953–961.

Rehman, I. H., A. Kar, R. Raven, D. Singh, J. Tiwari, R. Jha, P. K. Sinha, and A. Mirza. 2010. Rural energy transitions in developing countries: A case of the Uttam Urja initiative in India. *Environmental Science & Policy* 13(4): 303–311.

Roy, J. 2010. *Iron and Steel sectoral approaches to the mitigation of climate change: Perform achieve and trade in India*. Working Paper. Climate Strategies. http://www.climatestrategies.org/research/our-reports/category/54/274.html. (accessed October 20, 2014).

Roy J., S. Dasgupta, and D. Chakrabarty. 2017. Deep decarbonisation in industries: What does it mean for India? In *Realising long-term transitions towards low carbon societies:Impulses from the 8th Annual Meeting of the International Research Network for Low Carbon Societies*, S. Lechtenböhmer and K. Knoop (Eds.), 53–91. Wuppertal: Wuppertal Spezial 53, Wuppertal Institut für Klima, Umwelt, Energie. http://nbn-resolving.de/urn:nbn:de:bsz:wup4-opus-66360.

Sanderson, B. M., B. C. O'Neill, and C. Tebaldi. 2016. What would it take to achieve the Paris temperature targets? *Geophysical Research Letters* 43(13):7133–7142.

Times of India (TOI). 2014. Experts welcome e-rickshaw guidelines. New Delhi: The Times of India.

United Nations Framework Convention on Climate Change (UNFCCC). 2015. Paris agreement. http://unfccc.int/paris_agreement/items/9485.php. (accessed March 20, 2017).

Wall Street Journal. 2014. A nightmare death renews controversy over Delhi's electric rickshaws. New Delhi: The Wall Street Journal, India.

9

Bridges and Barriers for Governance of Water Resources in South Asia—Insights from Brahmaputra Basin in India

Navarun Varma and Sanjoy Hazarika

CONTENTS

9.1 Water in South Asia—Resource or Challenge? ... 161
9.2 Brahmaputra—A Case of "Misfit" River ... 166
  9.2.1 Misplacing Trust While "Fitting" the River ... 166
  9.2.2 Creating a Different Beast While "Fitting" the River ... 169
  9.2.3 Need to Create "Misfit" Knowledge ... 171
  9.2.4 Case of "Fit" and Challenge of Sustainability! ... 173
9.3 Barriers and Bridges ... 174
  9.3.1 Knowledge Gap ... 174
  9.3.2 Windows of Opportunity ... 175
9.4 Conclusion ... 177
References ... 178

9.1 Water in South Asia—Resource or Challenge?

The South Asian region (comprising Afghanistan, Bangladesh, Bhutan, China, India, Nepal, and Pakistan) has only 5 percent of the Earth's land area but must feed one-fourth (1.6 billion people) of the world's population (Rasul 2015). Economic growth and the scale of urbanization and industrialization have been impressive in most of the countries in this region in the last decade. Yet, 40 percent of the world's poor (population living below 1.25 United States dollars per day) and 35 percent of the world's undernourished still live in this region, and more than 56 percent of world's low birth-weight babies are born here. Poverty in the region is a paradox given abundance of natural resources, especially water, which is critical for food and energy production (Biswas 2008).

Water is indeed the most important piece of the puzzle. In terms of natural abundance of water, South Asia is unparalleled, given the flows of the Indus, Ganga, and Brahmaputra; the annual monsoon rains; and the vast reserves of fresh water in the snow-peaks and glaciers of the Hindu Kush Himalayas (HKH) (Wirsing 2008). The annual run-off of one of the most important interlinked basin systems of the region, that is, Ganga-Brahmaputra-Meghna (Figure 9.1), is around 1,150 bcm. The water availability in this interlinked system alone, considering only the surface water, is 3 times the water availability for the world (Biswas 2008; Bandopadhyay and Ghosh 2009). However, the entire region suffers from an issue of "too much or too little water" given its abundance during the monsoons but scarcity in dry season. For instance, the per capita availability of water resources in India has declined by approximately 60 percent over the last 50 years while seasonal scarcity has always been a challenge for Bangladesh given its minimal storage capacity because of its flat topography (Wirsing 2008). Quite contrasting is the monsoon scenario and its waves of flood events, with the latest and one of the worst in 2017 killing more than 1,200 people and devastating around 40 million people in India, Bangladesh, and Nepal (The Guardian 2017). The management of such major river systems is crucial for provisioning water, energy, and food to the current growing population (projected to reach 2.3 billion by 2050) while also securing ecosystem services to meet the needs of future generations (Rasul 2015). Although efficient management of water resource offers an entry point for sustainable development, the transboundary nature of water resources creates challenges due to complexities arising from varying national interests, policy, and institutional frameworks and epistemic boundaries (e.g., engineering, ecology, and economics) (Biswas 2008; Bandopadhyay 2009).

Scientific explorations of impacts of global climate change on South Asian water resources continue are in a variety of contexts. There is agreement among experts that there will be a gradual increase in temperature and a slight increase in total precipitation in the source areas of the Indus, Ganges, and Brahmaputra rivers by the end of this century (Immerzeel et al. 2013). Throughout the HKH region, glaciers may recede but runoff from the melting glaciers will keep rising until at least 2050 (Immerzeel et al. 2013). In this scenario, together with increasing precipitation, the total water availability may not decline during this century (Immerzeel et al. 2013). However, the changes in glacier volume and surface area can impact atmospheric circulation, which in turn influence the South Asian summer monsoon (Sun et al. 2010). There is still uncertainty about the impact of the combination of changes in various climatic factors on the Himalaya (for example, the impact of increasing greenhouse gases on a global scale vis-à-vis a short-lived climate pollutant such as black carbon from regional and local sources). Similar is the case with the variations in timing of precipitation, runoff, and frequency of extreme weather events as result of increasing temperatures and changing precipitation patterns. However, flooding

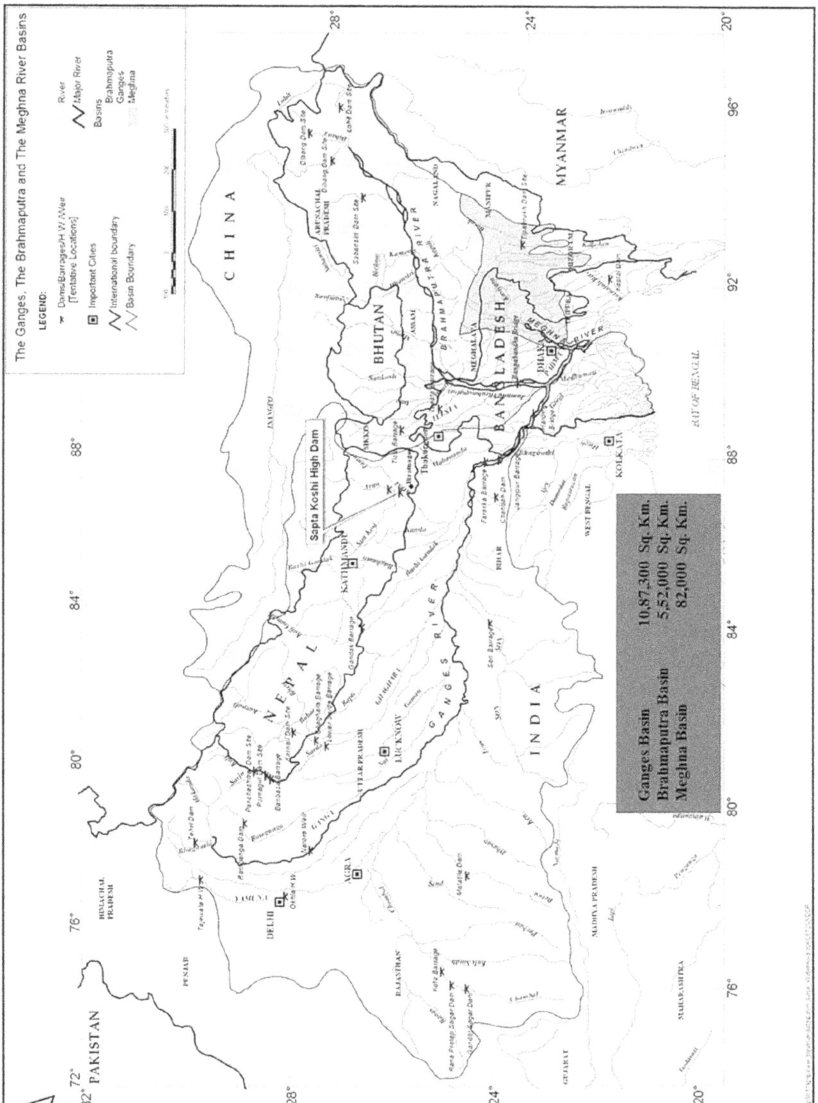

FIGURE 9.1
Map depicting the Ganga, Brahmaputra, and Meghna river basins.

events already appear to affect the Ganges-Brahmaputra-Meghna basin more frequently (Goswami et al. 2006; Bookhagen 2010), although indications of change in the frequency of droughts are still ambiguous (Burke and Brown 2008).

Given the risks of climate change, demand and supply of fresh water will be uncertain in the future. This uncertainty compounds the challenges for water resource management. While sustainable development is a worldwide challenge, the Eastern Himalaya, Tibetan plateau, and the Ganges-Brahmaputra-Meghna basin are likely to suffer extreme stress because of impacts from a set of critical interconnected social-ecological factors (Sud et al. 2015). Although the entire South Asia region has not met its Millennium Development Goals, the ongoing planning to adapt to Climate Change,[1] contribute to its mitigation, and achieve the Sustainable Development Goals (SDGs)[2] provide a fresh opportunity to examine the provisioning of water, energy, and food security. In addition, the implementation of the Sendai Framework for Disaster Risk Reduction (DRR) aims for substantial reduction of disaster risks and increased community resilience to future hazards (Selmi et al. 2015). Thus, there is a need to look at challenges and opportunities regarding water resources management with a fresh perspective for advancing climate change adaptation and achieving the SDGs and DRR goals.

There is no dearth of public policy and research focus on ecologically fragile and conflict-prone river basins of South Asia. Although there are success stories in different governance levels (e.g., sharing of disaster related information across South Asian Association for Regional Cooperation (SAARC) nations, community early warning systems, spring rejuvenation), sustainability of such policies and development practices has been a concern. We argue that there is a need for capacity among the water resource management community of practice in South Asia for developing systemic understanding of the dynamic complexity between river basin ecosystems, provisionary sectors (e.g., water, energy, and food), culture, and human aspirations. Years of understanding and managing river basins through reductionist schemas[3]

[1] For example, revision processes of Missions under National Adaptation Plan for Climate Change, State Action Plans for Climate Change, Intended Nationally Determined Contributions.

[2] For example, zero poverty (SDG 1), ending hunger and food insecurity (SDG 2), ensuring water security (SDG 6), access to modern energy (SDG 7), sustainable economic growth (SDG 8), sustainable consumption and production (SDG 12), and conservation, protection, and sustainable use of marine and terrestrial resources and ecosystems (SDGs 14 and 15) are interlinked and crucial for the region.

[3] "Human systems are complex and adaptive. An adaptive system operates by gathering information about its environment and its own interaction with that environment, identifying regularities in that information, condensing those regularities into a kind of "schema" or model, and acting in the real world on the basis of that schema." (Newell et al. 2005).

(e.g., through government departments of water resources, land revenue and so on, as well as disciplinary expertise of engineering, ecology, and so on) have led to consequences not often ecologically desirable and/or socially acceptable. This chapter uses the conceptual metaphor of institutional "fit," to illustrate the mismatches among public policy, discourses influencing problem framing, and social-ecological dynamics in the Brahmaputra River basin (BRB) of the Ganga-Brahmaputra-Meghna interlinked basin. We use the metaphor "fit" as currently used in sustainability science to understand the congruence or lack of it among design and operational scales of human institutions vis-à-vis ecological scales (e.g., spatial, temporal, and functional), social scales (e.g., social context in which rules will be embedded, inclusivity of rule making process to represent resources, and capacities of diverse stakeholder groups and opportunities of learning), and social-ecological systems (e.g., institutional design with empathy to context and its interplay with ecosystem functioning for improved social and ecological outcomes) (Epstein et al. 2015).

In this chapter we point out that scholars of Integrated Water Resource Management (IWRM)[4] have explored issues of "fit" for implementation or lack of it in the discourse in South Asian contexts. But there is silence regarding the role of social learning[5] for water resource management in such contexts while the concept of dialogue within IWRM provides space for its exploration. The chapter concludes with a proposition that current practices of dialogue (e.g., Track III dialogue processes for transboundary river basin management), when aided with transdisciplinary praxis,[6] can offer opportunities for shared understanding among diverse stakeholders, and initiate collaborations and co-design of solutions and management pathways within an iterative learning process.

[4] A discourse which takes geo-hydrological boundaries for understanding causes of land and water degradation and guiding public policy on water resources and rural development. Many argue that developing countries may benefit by drawing lessons from the IWRM experience in developed countries but there are also arguments on its neglect of institutional and epistemological legacies and dimension of power in public participation (Kurian 2004).

[5] Learning in multi-actor settings and also societal learning (i.e., structural changes within governance regimes). Changes can be triggered because of learning within strategies, assumptions that help in framing and finally values that influence such framing, in the governance regime of any context (i.e., the governance structure, legal and organizational frameworks as well as the cultural and socio-economic environment) and natural environment of the river basin (Pahl-Wostl 2007).

[6] To design effective action among individuals who have different disciplinary backgrounds. "Trans" means across, or "meta" and thus trans-disciplinary is either a "meta-discipline" or a form of practice that crosses disciplines (Ison 2008).

9.2 Brahmaputra—A Case of "Misfit" River

Half of the water available in the Ganges-Brahmaputra-Meghna interlinked basin comes from Brahmaputra. The Brahmaputra River basin (BRB) is a vast area covering some of the major sources of fresh water in the world and among the most complex, and politically, geographically, and ethnically diverse places on Earth. The basin (Figure 9.2) has a total population of more than 80 million people unevenly distributed among the riparian nations with the highest density around 828 persons per km^2 in the lowest riparian area of Bangladesh, followed by 143 persons per km^2 in India, 26 persons per km^2 in Bhutan, and the least population in uppermost riparian at 6 persons per km^2 in Tibet (Goswami 2008). In India, the basin area covers five of the eight states of the North-East Region (NER) that is Arunachal Pradesh, Assam, Meghalaya, Nagaland, and Sikkim. It is an abode to 200 of the 350 ethnic communities of the NER, each with its own characteristic culture and history that often reflects transboundary influence and geographical space which may not always match with the political boundaries within the NER (Goswami 2008). The NER has around 98 percent of its land borders with the other riparian nations of the Brahmaputra and also with Myanmar, and 2 percent with India that is the "chicken neck" near the state of West Bengal through which commodity exchange happens with rest of India. Thus, studies often refer to the whole NER with just 3 percent of the national population as "an anthropologist's delight and an administrator's nightmare" (Hazarika 2009).

The BRB has seen a range of complex contestations at the regional, national, and local levels (i.e., states/districts) over issues related to hydro-meteorological disasters, water sharing, hydro power projects, and population displacement as well as migration (Goswami 2003; Gohain 2008; Das et al. 2009; Hazarika 2009; Lahiri and Borgohain 2011; Baruah 2012; Varma and Mishra 2017). Here we present three interconnected issues of the BRB, that is, trust deficit, lack of innovations in policy and research, and unsustainable policy interventions and explain them through the metaphor of "fit" and illustrative examples.

9.2.1 Misplacing Trust While "Fitting" the River

At the transboundary level of the BRB, there is ambiguity about Chinese ambitions especially in their construction of hydropower projects. Almost as a competition for first user rights, India sanctioned as many as 168 dams in Arunachal Pradesh since the first decade of the twenty-first century before completion of the Zangmu hydropower station in the Tibet region. In 2001, the Brahmaputra basin was top ranked for potential hydroelectric schemes by the Central Electricity Authority (CEA) and in due course the government envisioned schemes for mega dams in the NER without much flood protection provisioning (Menon et al. 2003). It is noteworthy to mention that the Yarlung Tsango

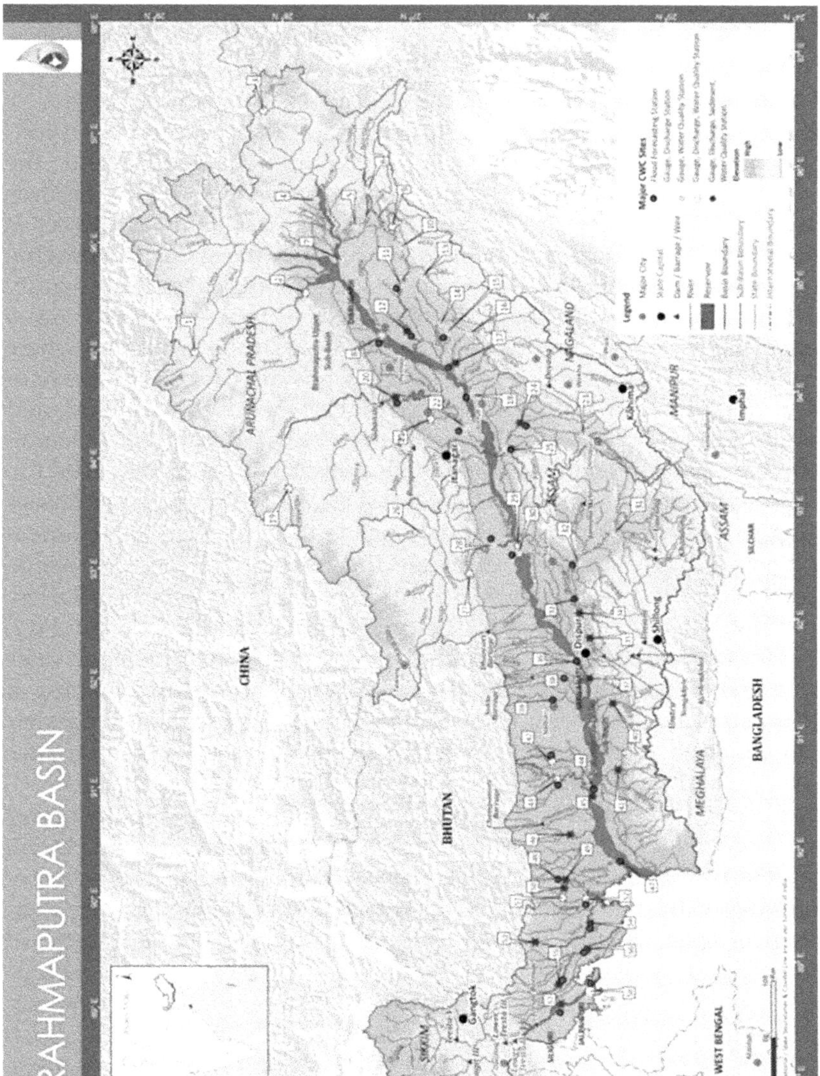

FIGURE 9.2
Map of the BRB with drainage and sub-basins.

(in Tibet Autonomous Region of China)-Brahmaputra (in North-East Region of India) region though an ecologically fragile and biodiversity hotspot lies in the periphery of the political space (polity) both for Beijing and New Delhi (e.g., Baruah 2005; Potter 2005). Since the independence of India, there exists a trust deficit between the ethnic communities of NER and the bureaucratic machinery of New Delhi, the genesis of which many find in the subjugation of tribal societies of the border areas under Nehruvian nation building and strengthened after the Sino-India war of 1962 (e.g., Baruah 2005).

As debates for the need of multipurpose dams for irrigation and flood control in NER versus plausible ecosystem impacts continued since the 1950s, the government sanctioned many of the river dams since CEA ranking (Biswas 2008; Baruah 2012; Varma and Mishra 2017). The hydropower project is complete in only one out of these sanctioned dams in the state of Arunachal Pradesh in Eastern Himalayas, that is, Ranganadi, while there is ambiguity over the future of the rest of the projects because of three major reasons: a huge financial crunch, reluctance by private players to invest in an area where land acquisition is tedious and unsure (many lands are caught in a maze of ownership ranging from private to collective, clans, and government), and extensive protests by civil society organizations (e.g., Baruah 2012; Duarah 2017; Rahman 2014). Even in the Ranganadi project, severe issues related to upstream-downstream linkages (diversion of flows within two tributaries originating in Arunachal-Dikrong and Ranganadi, deforestation, sediment load, and so on) have emerged. Although direct attribution to construction activities is difficult, a drastic increase in river bed level, rate of siltation, increasing water levels in flood as well as dry seasons, and extended inundation periods is observed in downstream Dikrong, while reduced flow rates and drying up of feeder channels to wetlands during lean season, lowering of groundwater tables, heavy sediment deposition, increase in braided channels during flood season, and erosion is observed in downstream Ranganadi since the project's operationalization (Vagholikar and Das 2010; Rahman 2014).

Another classic example is that of the currently stalled project of the Lower Subansiri dam in one of the major tributaries of Brahmaputra, the Subansiri. The attempts to legitimize this project range from lowering the scope of the Environment Impact Assessment in terms of stakeholder participation and ecological scale, labeling of risks of seismic activities, loss of traditional livelihoods raised by local academia as "myths," and incidences of corruption to get buy-in of downstream communities to stereotyping of mobilizations against it as "left-wing extremism" (Baruah 2012; Rahman 2014). Such attempts indicate the political will to "fit" the flows of a river in the polity's periphery for demands, that is, energy of a political economy at the center (Varma and Mishra 2017). In NER, where even construction of reservoirs for flood control are debatable given the high risks of seismic activities, this kind of hydro-development under conditions of sparse hydrological information and future uncertainty increases vulnerability of communities to disasters (Bandopadhyay 2009; Mahanta 2010; Baruah 2012). Thus, a development paradigm which does not pay attention to the perceived risks increases the vulnerability of already marginalized communities and reinforces the existing mistrust between New Delhi technocrats and the NER ethnic groups (Baruah 2012; Varma et al. 2016; Rahman 2017; Varma and Mishra 2017).

9.2.2 Creating a Different Beast While "Fitting" the River

In the last 30–40 years, the Brahmaputra River has widened by 50 percent by eroding upstream areas of its valley. This valley, with a total size of 56, 274 km^2 that comprises the state of Assam in NER of India, loses an average of 8,000 ha of land in the erosion process. In addition, Assam faces a staggering loss of property, infrastructure, farm, livestock, and lives in floods every year. Based on data collected at the village level and compiled by the Indian Central Water Commission, floods between 1953 and 2011 affected about 11 percent of the total area of the state of Assam, caused mean annual economic damage USD $56.3 \pm 5.43 \times 10^6$, and killed a total of 2,761 people with the highest toll of 497 in 2004. The official data does not record deaths caused by indirect causes like diseases such as malaria in areas of stagnant water (Wasson et al. 2017).

The most important policy intervention has been one of earthen embankment construction and continued maintenance since the enactment of the Assam Embankment and Drainage Act of 1954. Embankments were not new to Assam as they had been part of public policy since the pre-colonial Ahom dynasty rule (1228–1826) for controlling the quantity of water in farms to introduce wet rice cultivation for increasing agricultural production (Gogoi 2002). By the 1930s, the colonial need for cultivation and increasing production of cash crops like jute and thereafter the construction of railway tracks in flood plains reinforced the construction of embankments. But some of the then-technocrats recommended a cautious approach to adopt embankment construction for flood protection. They based their arguments on evidences of ponding and swamps that were created by old embankments of sixteenth and seventeenth centuries and also on the risk of disastrous flood that embankments can lead, through sedimentation of river channel between embankments, to raising flood levels reaching a floodplain higher than the original level and also resulting in an ever-increasing need for higher and stronger embankments.[7] This is now conceptualized as one of the dimensions of "levee effect"[8] (Di Baldassarre 2015; Wasson et al. 2017).

[7] Also, in personal communications with Dr. Arup Jyoti Saikia, Department of Humanities and Social Sc., Indian Inst. of Tech. (IIT), Guwahati, Assam.

[8] Flood protections like dams and levees may increase flood losses because they lead to new development in the floodplain, which incurs catastrophic losses when such flood protections fail. This phenomenon is known as the "levee effect." It is a classic example to illustrate how short-term risk reduction strategy can lead to long-term vulnerability. Levees or embankments do reduce damage first which reduces perceived risks and initiates development in flood plains thus increasing vulnerability and eventually when they are overtopped (or breached as in case of Brahmaputra) large losses are incurred (Kates et al. 2006).

But the turning point came after Indian independence when the continued floods until 1955 after the 1950 earthquake (measuring 8.3 Richter scale) damaged crops worth about USD 2.1 × 10^6 in 1954, that is, USD 7.7 × 10^6 in 2017, destroyed 65 percent of paddy, 53 percent of jute, and inundated almost 40 percent of the state, affecting about 1.2 million people (Wasson et al. 2017). Influenced by China's flood protection with embankments and legitimized here by the Assam Embankment and Drainage Act, 1953, the government constructed 4,000 km of embankments in Assam along with at least 700 km of drainage channels by 1978. Breaches had occurred from very early on and within a year in few of the new constructed embankments, but soon a political economy favoring a network of bureaucrat-politician and contractor developed, which still motivates policy towards embankment construction and associated maintenance (raising and strengthening) in the twenty-first century. At the same time a dependency for protection has emerged on embankments from agrarian communities (shifting from a "slash and burn" culture to settled agriculture) and even townships (with oil installations and plantations) that have slowly prospered in flood plains as perceptions of flood risk reduced with more embankments, thus increasing the potential loss and damage from flood events, yet another important dimension of the "levee effect" (Hazarika 2009; Di Baldassarre 2015; Varma and Mishra 2017; Wasson et al. 2017).

The analysis of Wasson et al. (2017) on available information on flood damages and loss of lives during floods in Brahmaputra Valley from 1956 to 2011 suggests a situation that is consistent with the concept of the levee effect from social hydrology and some of the early apprehensions on the paradigm of flood protection through embankments. There is a relation between flood discharge and damage during the period 1956–1981, indicating damage as a function of discharge and exposure as most of the floodplain development happened during this phase. While there is no significant relation between 1982 and 2011, a phase when most of the embankments were complete. During the second phase, high damage years did not always coincide with years of high flood discharge supporting "levee effect" which creates a few large damage totals but no change or a decline in overall damage. The analysis of the death toll in both phases indicated a positive relation between discharge and death, with higher significance in the latter phase. This analysis illustrates that embankments had no direct impact on the death toll. But with more embankments came more exposure to floods by floodplain development along with breaches (e.g., Varma and Mishra 2017), thereby increasing deaths. Also, deaths were high as discharges were high for both phases, once again illustrating that embankments do not reduce deaths but may increase them indirectly (Wasson et al. 2017). Thus, a "different beast" is created while "fitting the river" first for increasing productivity and connectivity for an imperial agenda and then for reducing disaster risk post-independence and finally now as if in a state of technological lock-in without alternatives.[9]

9.2.3 Need to Create "Misfit" Knowledge

Contemporary studies including NER studies have separated the issues of the BRB into three different debates within their respective epistemic boundaries. (1) The "fluvial geomorphology" framework draws from the disciplines of geology and hydrology while explaining the changes in flow pattern in different reaches of the river basin and dynamics of sediment transport in different time periods (Goswami and Das 2003; Sarma 2005). (2) The political economy framework, drawing from the disciplines of history and economics, explains the different interventions in the BRB as influenced by evolution of different resource appropriation regimes from pre-colonial to postcolonial times (Das and Saikia 2011; Saikia 2012; Saikia 2014). (3) The discipline of political science explains the course of defining political space within the BRB and the resulting struggle for identities among ethnic communities (Phukon and Dutta 1997; Baruah 2005). These conceptual frameworks have been part of different discourses that have shaping public opinion and policy towards disaster and governance issues in the context. The categorization of the frameworks follows (Varma 2016):

- Formal policy discourse of flood control and disaster management in Assam often uses elements of the fluvial geomorphology framework (Flood Report of WRD, GoA 1955–2009; Assam, National Disaster Risk Reduction Portal, NIDM).
- Academia and civil society draw on elements of the political economy by debating about the legitimacy of hydropower development projects and arguing for strengthening the EIA studies (Menon et al. 2003; Baruah 2012).
- Analysis of the relations between Indian polity and citizens of NER uses identity and space politics conceptualization and proposes better regional cooperation with Southeast Asian nations for targeting the biases among the communities of the region and the nation (Baruah 2005).

However, such knowledge falls short in seeking solutions to current paradoxes of water resources management in the BRB. On one hand, *de-novo* institutions for engaging with society and developing capacities for flood

[9] In personal communication with Prof. Robert J. Wasson, Senior Research Fellow, Inst. of Water Pol, Lee Kuan Yew School of Pub. Pol. (LKYSPP), National Univ. of Singapore (NUS).

disaster management (e.g., resource management, co-ordination, and community-based resilience) have been constituted,[10] the states of the NER shaped their State Action Plan(s) for Climate Change, counter insurgency operations have been largely successful especially in Assam (Baruah 2005), and there are plans for hydro-development in the NER (Menon et al. 2003). On the other hand Brahmaputra Valley seems to be in a state of "levee effect" while embankments and demands of relief packages from national government dominate the policy debate after every flood cycle, protests against dam construction by different ethnic student organizations continue along with serious questions by local academia and civil society on the credibility of EIAs for such projects and farmer agitations against land loss (Varma 2016; Varma and Mishra 2017; Wasson et al. 2017). Such paradoxes indicate a disconnect between public needs, aspirations, and public policy.

From the previous discussion, the search of a fit between the water flows, geo-political, and cognitive boundaries creates new issues, leads to emergence of a maladapted society to hazards, and reinforces mistrust across actors within and across nations. Such conditions also inhibit information sharing, create typifications, and increase vulnerability to future risks. There are ongoing experiments with controlled flooding and sedimentation through flood polders and livelihood experiments in *char* islands[11] (Zaman 2012) but the environment of mistrust[12] inhibits learning across borders and stakeholders. Because of lack of scope of experimentation and innovations in policy and practice, an entire generation of farming communities is not content with government interventions for flood control, dependent on flood relief, and associated with ethnic political mobilizations against development projects (Varma and Mishra 2017).

This, we argue, should be motivating research in the BRB towards a praxeology exploring issues not only through the existing influential frameworks separately, but also coupling two or more, and innovating methodologies to engage stakeholders in the research process. Such stakeholder engagements will not only illustrate rich experiences and tacit knowledge (e.g., Gayawali and Thompson 2016) but can also build trust and acceptance of solutions. There is a need of a "misfit" knowledge inquiry process which can be participatory in design and have the potential of problem solving. This is a capacity need among researchers and development practitioners working towards effective and efficient water resource

---

[10] For example, the Flood and River Erosion Management Agency of Assam (FREMAA) for managing funds from the Asian Development Bank (ADB), and the Assam Disaster Management Agency (ASDMA) for coordination capacities of government departments and community-based disaster resilience.

[11] Sand bars formed in river channel due to depositions of sediment by the river flow, a characteristic of braided rivers like Brahmaputra, is locally known as *char* in Assam in India and Bangladesh.

[12] For example regarding Teesta water sharing between India and Bangladesh (Biswas 2008).

management in conflict-prone and ecologically fragile river basins like the BRB across South Asia (Varma and Mishra 2017).

9.2.4 Case of "Fit" and Challenge of Sustainability!

As we see certain indigenous coping practices become obsolete given new kind of hazards (e.g., flash floods and coarse sediment deposition after floods) in the valley, yet there are contexts within the BRB where community mobilizations and partnerships have emerged which could tackle novel issues. For example, in the Indian NER state of Sikkim within Eastern Himalayas in Teesta sub-basin, certain policy-driven interventions have worked in favor of people. The best example is the *Dhara Vikas Yojana* (i.e., Spring Development Program) of the state's Rural Management Development Department (RMDD). The scheme generated resources for community engagement from the national Mahatma Gandhi Rural Guarantee Scheme (MGNREGA), created a scope for deliberation among resource users, experts, and government employees, and attracted additional funding form donor agencies for scaling. In addition, it has also trained and developed the capacity of community leaders as para-hydro geologists. After successful interventions in terms of spring discharge and water storage in South and East Sikkim, the government replicated this intervention in other NER states like Meghalaya where drying up of springs was emerging as an issue for mountain communities. Significantly, the intervention drew the attention of national level policy makers and inter-governmental agencies like International Centre for Integrated Mountain Development (ICIMOD) for replication.[13]

But even such successful intervention meets with sustainability challenge as it does not incorporate future risks and lacks a proper policy evaluation framework. Farm communities in the mountains of Sikkim are increasingly transforming their livelihoods from subsistence farming, pastoralism, and collection of timber/non-timber forest products to water intensive cash crop-based agriculture (e.g., cardamom) and tourism (home stays, treks and so on). Such a change has increased the pressure on already vulnerable resources like springs (e.g., through sprinkler irrigation, increasing demands from enterprises like home stays, and so on). Moreover, there is lack of attention to management issues of infrastructures (e.g., trenches for percolation of rainwater, spring discharge information, and so on) created during spring rejuvenation[14] and capacity building of local governance institutions

[13] RMDD had announced that the intervention revived 54 springs and farmers depending on these springs earn INR 330 USD to 500 USD per month which was not possible 5–6 years ago. Presented by Joint Secretary, MGNREGA, Government of Sikkim, in a workshop on Water Access and availability in mountain areas of Sikkim organized by Himalayan Adaptation, Water, and Resilience (HI-AWARE), (http://www.hi-aware.org/) on November 26, 2015.

[14] For example, trenches dug for proliferation of groundwater are away from sites of beneficiaries, that is, forest areas and up-hill villages, and hence there are no incentives for their maintenance.

and civil society organizations for maintenance and continuity; that is, the dimension of evaluation of policy output and outcome which is crucial for sustainability and scaling of the intervention (Sen et al., manuscript under review). Hence, although successful now, as were the embankments of Assam in early 1970s (e.g., Varma and Mishra 2017), there is a need for iteration in any ongoing program informed by proper policy evaluation involving stakeholders and learning from implementation process, outputs, and outcomes (e.g., Rijke et al. 2012; Pahl-Wostl et al. 2013). This brings us back to the discussion of the capacity need of a "misfit" knowledge inquiry process to inform public policy and practice.

9.3 Barriers and Bridges

9.3.1 Knowledge Gap

Development planning has proven to be myopic in the BRB, like many other South Asian contexts, as it rarely takes into the account complexity of river basin ecosystem, services flowing from it, and human dependencies. Over the years, governments in South Asia have developed an array of strategies trying to tackle some of these critical concerns. However, government institutions miss out on the intrinsic linkages within ecosystems, provisionary sectors (e.g., water, energy, and food), culture, and human aspirations due to tendency of working in reductionist schemas (e.g., departments, disciplines, and so on) (e.g., Sud et al. 2015). This situation often leads to ignorance of trade-offs and missed opportunities of synergies and co-benefits across ecological and governance scales which triggers policy failure, maladaptation, and reinforcing of the problem.

Many research and practice-oriented studies have recommended the need for the Integrated Water Resource Management (IWRM) in responding to current mismatches among demand and supply, issues of social-ecological sustainability, and climate impacts. But barriers for its application emerge owing to the transboundary character of the resource, more so in South Asia due to influence of a legacy of territorial disputes over public policy and constraints in evolution of institutions and epistemologies for dealing with inter-sectoral issues (e.g., land use planning, agriculture, disaster management, hydropower generation, ground-water management, and so on) (Shah et al. 2006; Haines 2014; Price and Mitra 2016; Barua et al. 2017). Studies of "fit" in water sector for implementation of IWRM in contexts of South Asia (e.g., Shah et al. 2006) have not yet explored the role of human agency in reproduction and/or change of rules and norms. There remains a dearth of literature in exploring the role or opportunity of social learning in water resources management in South Asia.

In several countries of the region there are instances of Track II diplomacy where international agencies, think tanks, and civil society have facilitated shared learning exercises of pressing issues (e.g., flood disaster in transboundary rivers like Koshi, Brahmaputra, and so on) and best practices (e.g., early warning systems, spring rejuvenation, and so on) across borders. Although they create an environment of dialogue and learning, the experience has been that translating the lessons into policy and practice requires political willingness, creativity, and leadership (Price and Mitra 2016).

Track II engagement was most active in the 1990s and in first decade of the 2000s and led to successful negotiations between India and Bhutan regarding Brahmaputra as well as India and Bangladesh regarding Ganga water. However, many other efforts are yet to take off, especially at the Track III and Track IV levels—civil society to civil society and civil society to community partly because of the intense mistrust between governments which spills over into other sectors.[15] Efforts by international organizations such as International Union for Conservation of Nature and Natural Resources (IUCN) to develop a middle ground level involving universities, professionals, students, government officials, and diplomats have generated interests and goodwill but are yet to get off the ground and have not been sustained. Thus, there has not been much opportunity for social learning. Given support and sustenance, can such democratic efforts succeed in a region mired in a legacy of mistrust? This remains a key and complex question.[16]

9.3.2 Windows of Opportunity

It is noteworthy to look into some of the current opportunities for operationalizing concept of "dialogue" from the Integrated Water Resource Management (IWRM) discourse that is already embedded in the ongoing process of decentralization of resource management in the entire South Asia region (e.g., Joint Forestry Management, Community based forestry, Water Users Association, and so on) (Shah and Prakash 2010). In regions like the BRB where there is a deadlock on Track I diplomacy due to competing national interests (China, India, and Bangladesh), several think tanks with support of international agencies have started dialogue forums and trainings on communication for senior and mid-level bureaucrats and non-government organizations (associated with water sector) in order to build capacity for Track III and also influencing Track II engagements in the region (e.g., SaciWATERS 2016; Price and Mitra 2016). It may be worthwhile to explore only this conceptual dimension of IWRM; as for operationalization of the full discourse, there is a need for multilateral treaties and often also for a river basin authority. Both needs may

[15] Travelers from Bangladesh, Pakistan, and China need special security clearance from New Delhi to visit India even for academic conferences.

[16] Second author's reflections from engagement.

be difficult to meet and may not be even desirable as there are arguments which suggest that such basin wide institutions may just reinforce rigidity and "technocratic control and command" because the dominant actors for water management in the region are still engineers and/or bureaucrats heavily dependent on the former for increasing production and supply (Shah and Prakash 2010). However, the dialogue has to be in multiple jurisdictional levels that is transnational (e.g., riparian nations sharing water resources of Brahmaputra), national-sub-national (e.g., the national government in Delhi with the state governments of Assam and Arunachal Pradesh when it comes to issues of the BRB) and beyond formal government agencies integrating civil society, media, and community representatives (e.g., the ongoing Brahmaputra Dialogue process, Barua et al. 2017).

Following from our discussion of need of "misfit" knowledge, it is necessary for a robust yet flexible methodology for any meaningful dialogue; one that can challenge worldviews, raise questions and new issues, and facilitate negotiations. Such a language for engagement blurs cognitive boundaries and typifications (e.g., stereotypes created through cultural legacies and/or boundaries created through formal institutional guidelines), demystifies experts' discourses (e.g., concepts of climate change vulnerability, disaster risk reductions, resilience, and so on), and learns nuances from stakeholders' experiences. We believe that it is about time to build capacity in transdisciplinary techniques, like participatory systems modeling, participatory geographic information system (GIS)-based tools, role playing/serious games, and so on (e.g., Voinov and Bousquet 2010) among the South Asian water resources management community of practice. Systems can use such techniques to design deliberations in current dialogue opportunities in the region like the Brahmaputra Dialogue (Barua et al. 2017).

The relevance of transdisciplinary praxis is most in conflict-prone areas like the BRB where mistrust runs deep among riparian nations as well as stakeholders within a single polity. The benefit of using such techniques is that they de-humanize an issue and illustrate the social constructs of a problem situation through models and maps. This technique may create empathy to the "other," facilitate identification of knowledge gaps, learning from different worldviews as well as experimentation (which can be through computer aided simulations and/or actual pilots), and can strike collaborations and initiate negotiated outcomes. However, the most important prerequisites of such methodology are facilitation skills, design of iterative learning processes, well-informed but unbiased stakeholder identification, and evaluation of the learning outcomes and process for improvement and sustainability (e.g., Vennix et al. 1996; Wester et al. 2003; Linnerooth-Bayer et al. 2006; Rist et al. 2007; Hall and O'Rourke 2014).

There is much research on the influence of social learning, mediated by transdisciplinary praxis (e.g., Pohl 2014), for shifts in governance paradigm in European river basins (e.g., Magnuszewski et al. 2005; Pahl-Wostl et al. 2013). Some transdisciplinary techniques (e.g., visualization tools used in the

Bangladesh delta plan 2100, http://www.climateadaptationservices.com/en/bangladesh-deltaplan-2100; Varma et al. 2015; Tschakert et al. 2016) have been used in climate adaptation and disaster risk reduction research and policy in South Asian contexts. However, there is still a gap in understanding as to how far such techniques can contribute in influencing public policy and development practice in regions of trust deficit and top-down governance. We believe that such a question demands an experimental research design and development of methodology as heuristics (Huutoniemi 2014). This demand means that the researcher and research team will have to work out innovations in research design (including choice of technique, methodological steps, design of workshop, and/or fieldwork, and so on) according to the problem context, balancing cultural sensitivities and scientific rigor. Logic and culture can inform studies using such methodology; and, hence, facilitate shared understanding of issues and motivate towards "desirable and feasible" solutions (Checkland and Scholes 1990).

9.4 Conclusion

Given the policy goals of sustainable development, climate change action and disaster risk reduction, it is urgent for fresh perspectives for water resources management in South Asia. Until now, there have been undesirable consequences while engineering a "fit" for water resources management in different contexts of South Asia and here we illustrate challenges due to "misfit" in the BRB. Existing epistemologies have also fallen short in resolving the management paradoxes that have emerged during this "misfit." However, "fit" may emerge through partnerships across government, civil society, academia, and community as in the case of spring rejuvenation policy intervention in Sikkim. But sustaining such practices depends on taking cognizance of social-ecological changes this further creating the need of a participatory approach for policy evaluation and iterations in the policy design.

This creates capacity needs for a transdisciplinary praxis among the South Asian water resources management community of practice. However, some of the same consequences of "misfit" like trust deficit may not provide the necessary social learning opportunities, for example, learning platforms and stakeholder collaborations, and rather may have facilitated typifications which may create pre-dispositions inhibiting any meaningful dialogue. But amid such "barriers," current enthusiasm for stakeholder engagements for water resources management, for example Track III engagement processes for river basin management like the Brahmaputra dialogue, offer themselves as "bridges" for impacting changes in policy paradigm and public opinion. It almost comes as a responsibility of the praxis community to use and customize transdisciplinary techniques to structure deliberations in such ongoing engagements for shared understanding of the issues and in facilitation for co-design of solutions.

References

Aitisi-Selmi, A., S. Egawa, H. Sasaki, C. Wannous, and V. Murray. 2015. The Sendai framework for disaster risk reduction: Renewing the global commitment to people's resilience, health, and well-being. *International Journal of Disaster Risk Science.* 6(2):164–176.

Bandyopadhyay, J. 2009. *Water, Ecosystems and Society – A Confluence of Disciplines.* New Delhi, India: SAGE Publications.

Bandopadhyay, J., and N. Ghosh. 2009. Holistic engineering and hydro-diplomacy in the Ganges-Brahmaputra-Meghna Basin. *Economic and Political Weekly* Vol. XLIV(45):50–60.

Barua, A., S. Vij, and M. Z. Rahman. 2017. Powering or sharing water in the Brahmaputra River basin. *International Journal of Water Resources Development* 1–15.

Baruah, S. 2005. *Durable Disorder: Understanding the Politics of Northeast India.* New Delhi: Oxford University Press.

Baruah, S. 2012. Whose river is it anyway? Political economy of hydropower in the Eastern-Himalayas. *Economic and Political Weekly* Vol. XLVII(29):41–52.

Biswas, A. K. 2008. Management of Ganges-Brahmaputra-Meghna system: Way forward. In *Management of Transboundary Rivers and Lakes. Water Resources Development and Management,* O. Varis, A. K. Biswas, and C. Tortajada (Eds.). Berlin, Germany: Springer.

Bookhagen, B., and D. W. Burbank. 2010. Toward a complete Himalayan hydrological budget: Spatiotemporal distribution of snowmelt and rainfall and their impact on river discharge. *Journal of Geophysical Research* 115:1–25.

Burke, E. J., and S. J. Brown. 2008. Evaluating uncertainties in the projection of future drought. *Journal of Hydrometeorology* 9:292–299.

Checkland, P., and J. Scholes. 1990. *Soft Systems Methodology in Action.* West Sussex, UK: Wiley.

Das, P. J., D. Chutiya, and N. Hazarik. 2009. Adjusting to floods on the Brahmaputra plains, Assam, India. Aranayak, Assam: International Centre for Integrated Mountain Development (ICIMOD).

Das, D., and A. Saikia. 2011. Early twentieth century agrarian Assam: A brief and preliminary overview. *Economic and Political Weekly* XLVI(41):73–80.

Di Baldassarre, G., A. Viglione, G. Carr, L. Kuil, K. Yan, L. Brandimarte, and G. Bloschl. 2015. Debates-Perspectives on socio-hydrology: Capturing feedbacks between physical and social processes. *Water Resources Research* 51:4770–4781.

Duarah, C. K. 2017. Protests against dams on the Siang continue in Arunachal Pradesh. *The Quint.* Available online at https://www.thequint.com/news/environment/protests-against-dams-on-the-siang-continue-in-arunachal-pradesh (Accessed July 16, 2018).

Epstein, G., J. Pittman, S. M. Alexander, S. Berdej, T. Dyck, U. Kreitmair, K. J. Rathwell, S. Sergio Villamayor-Tomas, J. Vogt, and D. Armitage. 2015. Institutional fit and the sustainability of social–ecological systems. *Current Opinion in Environment Sustainability* 14:34–40.

Gayawali, D., and M. Thompson. 2016. Restoring development Dharma with Toad's Eye science? In: *States, Markets and Society-New Relationships for a New Development Era*, M. Melch (Ed.). *IDS Bulletin* 47(2A).

Gogoi (Nath), J. 2002. *Agrarian System of Mediaeval Assam*. New Delhi, India: Concept Publishing Company.

Gohain, H. 2008. Big dams, big floods: On predatory development. *Economic and Political Weekly* 43(30):19–21.

Goswami, D. 2003. Flood problem of Assam: Causes and remedies. In *Flood Problem of Assam: Causes and Remedies*, P. C. Sabhapandit (Ed.). New Delhi, India: Omsons Publications.

Goswami, D. 2008. Managing the wealth and woes of the river Brahmaputra. *Facets of North-East. Ishani* 2(4):1–11.

Goswami, B. N., V. Venugopal, D. Sengupta, M. Madhusoodanan, and P. K. Xavier. 2006. Increasing trend of extreme rain events over India in a warming environment. *Science* 314:1442–1445.

Haines, D. 2014. Disputed rivers: Sovereignty, territory and state-making in South Asia, 1948–1951. *Geopolitics* 19(3):632–655.

Hall, T. E., and M. O'Rourke. 2014. Responding to communication challenge in transdisciplinary sustainability science. In *Transdisciplinary Sustainability Studies: A Heuristic Approach*, K. Huutoniemi and P. Tapio (Eds.). New York: Routledge.

Hazarika, S. 2009. Ethnic conflict and civil society in the North-East. *Blood Sport*, 5 and 6. Available online at http://www.littlemag.com/bloodsport/sanjoyhazarika.html (Accessed July 16, 2018).

Huutoniemi, K. 2014. Introduction: Sustainability, transdisciplinarity and the complexity of knowing. In *Transdisciplinary Sustainability Studies: A Heuristic Approach*, K. Huutoniemi and P. Tapio (Eds.). New York: Routledge.

Immerzeel, W. W., F. Pellicciotti, and M. F. P. Bierkens. 2013. Rising river flows throughout the twenty first century in two Himalayan glacierized watersheds. *Nature Geoscience* 6:742–745.

Immerzeel, W., L. van Beek, M. Konz, A. B. Shrestha, and M. Bierkens. 2012. Hydrological response to climate change in a glacierized catchment in the Himalayas. *Climatic Change* 110:721–736.

Ison, R. L. 2008. Methodological challenges of transdisciplinary research: Some systemic reflections. *Nature Sciences Societies* 16:241–251.

Kates, R. W., C. E. Colten, S. Laska, and S. P. Leatherman. 2006. Reconstruction of New Orleans after Hurricane Katrina: A research perspective. *Proceedings of the National Academy of Sciences* 103(40):14653–14660.

Kurian, M. 2004. *Institutions for Integrated Water-Resources Management in river basins: An analytical framework*. Working Paper 78. Sri Lanka: International Water Management Institute.

Lahiri, S. K., and J. Borgohain. 2011. Rohmoria's challenge: Natural disasters, popular protests and state apathy. *Economic and Political Weekly* XLVI(2):31–35.

Linnerooth-Bayer, J., A. Vari, and M. Thompson. 2006. Floods and fairness in Hungary. In *Clumsy Solutions for a Complex World: Governance, Politics and Plural Perception*, M. Verweij and M. Thompson (Eds.). New York: Palgrave Macmillan.

Magnuszewski, P., J. Sendzimir, and J. Kronenberg. 2005. Conceptual modeling for adaptive environment assessment and management in Barycz. *International Journal of Environmental Research and Public Health* 2(2):194–203.

Mahanta, C. 2010. India's North East and hydropower development: Future security challenges. *South Asian Survey* 17(1):131–146.

Menon, M., N. Vagholikar, K. Kohli, and A. Fernandes. 2003. Large dams in the North East: A bright future. *The Ecologist Asia* 11(1):3–8.

Newell, B., C. L. Crumley, N. Hasan, E. F. Lambin, C. Pahl-Wostl, A. Underdal, and R. Wasson. 2005. A conceptual template for integrative human-environment research. Global *Environment Change* 15:299–307.

Pahl-Wostl, C., G. Becker, C. Knieper, and J. Sendzimir. 2013. How multilevel societal learning process facilitate transformative change: A comparative case study analysis on flood management. *Ecology and Society* 18(4):58.

Pahl-Wostl, C., M. Craps, A. Dewulf, E. Mostert, D. Tabara, and T. Taillieu. 2007. Social learning and water resources management. *Ecology and Society* 12(2):5.

Phukon, G. and N. L. Dutta. 1997. *Politics of Identity and Nation Building in Northeast India*. New Delhi, India: South Asian Publishers.

Pohl, C. 2014. From complexity to solvability: The praxeology of transdisciplinary research. In *Transdisciplinary Sustainability Studies: A Heuristics Approach*, K. Huutoniemi and P. Tapio (Eds.). New York: Routledge.

Potter, P. 2005. Governance of China's periphery: Balancing local autonomy and national unity. *Columbia Journal of Asian Law* 19(1):293–322.

Price, G., and S. Mittra. 2016. Water, ecosystems and energy in South Asia. Making cross-border collaboration work. *Chatham House Report*.

Rahman, M. 2014. *Territory, tribes, turbines: Local community perceptions and responses to infrastructure development along the Sino-Indian Border in Arunachal Pradesh*. Occasional Paper. No. 7. New Delhi, India: The Institute of Chinese Studies.

Rasul, G. 2015. Food, water, and energy security in South Asia: A nexus perspective from the Hindu Kush Himalayan region. *Environmental Science and Policy* 39:35–48.

Rijke, J., R. Brown, C. Zevenbergen, R. Ashley, M. Farrelly, P. Morison, and S. van Herk. 2012. Fit for purpose governance: A framework to make adaptive governance operational. *Environmental Science and Policy* 22:73–84.

Rist, S., M. Chindambaranthan, C. Escobar, U. Weismann, and A. Zimmermann. 2007. Moving from sustainable management to sustainable governance of natural resources: The role of social learning processes in rural India, Bolivia and Mali. *Journal of Rural Development* 23:23–37.

Saikia, A. 2012. The historical geography of the Assam violence. *Economic and Political Weekly* Vol. XLVII(41):15–18.

Saikia, A. 2014. *A Century of Protests: Peasant Politics in Assam Since 1900*. New Delhi, India: Cannaught Place.

Sarma, J. N. 2005. Fluvial process and morphology of the Brahmaputra River in Assam, India. *Geomorphology* 70:226–256.

Sen, S. M., A. Singh, N. Varma, D. Sharma, and A. Kansal. (manuscript under review). Analysing social networks to examine the changing governance structure of springsheds: A case study of Sikkim in the Indian Himalayas. *Environmental Management*.

Shah, T., I. Makin, and R. Sakthivadivel. 2006. Limits to leapfrogging: Issues in transposing successful river basin management institutions in the developing world. In *Integrated Water Resources Management, Global Theory, Emerging Practice and Local Needs*, P. P. Mollinga, A. Dixit, and K. Athukorala (Eds.). New Delhi, India: Sage Publication.

Shah, A., and A. Prakash. 2010. In *Integrated water resource management in India: From critique to constructive engagement*, 333–346. Hyderabad, India: Central Research Institute for Dryland Agriculture.

Siddique, H. 2017. South Asia floods kill 1,200 and shut 1.8 million children out of school. *The Guardian*. Accessed online: https://www.theguardian.com/world/2017/aug/30/mumbai-paralysed-by-floods-as-india-and-region-hit-by-worst-monsoon-rains-in-years.

The South Asia Consortium for Interdisciplinary Water Resources Studies (SaciWATERS). 2016. Role of dialogue in transboundary water management-The Brahmaputra basin. Brahmaputra dialogue. Policy Brief. Issue no. 1. Transnational policy dialogue for improved water governance of the Brahmaputra River-Phase III.

Sud, R., A. Mishra, N. Varma, and S. Bhadwal. 2015. Adaptation policy and practice in densely populated glacier-fed river basins of South Asia: A systematic review. *Journal of Regional Environmental Change* 15(5):825–836.

Sun, Y., Y. Ding, and A. Dai. 2010. Changing links between South Asian summer monsoon circulation and tropospheric land-sea thermal contrasts under a warming scenario. *Geophysical Research Letters*. 37(2):L02704.

Tschakert, P., P. J. Das, N. S. Pradhan, M. Machado, A. Lamadrid, M. Buragohain, and A. Hazarika. 2016. Micropolitics in collective learning spaces for adaptive decision making. *Global Environment Change* 40:182–194.

Vagholikar, N., and P. Das. 2010. Damming the North-East India-juggernaut of hydropower projects threatens social and environmental security of region. *Articles, India Water Portal*. http://www.indiawaterportal.org.

Varma, N. 2016. Disaster and Governance in Brahmaputra basin of India: Case study of an ecological surprise within Assam. Thesis-Ph.D. *TERI School of Advanced Studies*. Abstract available online at http://lib.icimod.org/record/32739.

Varma, N., U. Kelkar, S. Bhardwaj, P. Singh, and A. Mishra. 2015. Climate change, disasters and development: Testing the waters for adaptive governance in India. *Vision: Journal of Business Perspective: Special Issue on Governance and Public Policy* 18(4):327–338.

Varma, N., and A. Mishra. 2017. Discourses, narratives and purposeful action-Unravelling the social-ecological complexity within Brahmaputra Basin in India. *Journal of Environmental Policy and Governance* 27(3):207–228.

Vennix, J. A. M. 1996. *Group Model Building. Facilitating Team Learning Using System Dynamics*. West Sussex, UK: Wiley.

Voinov, A., and F. Bousquet, 2010. Modeling with stakeholders. *Environment Modeling and Software* 25:1268–1281.

Wasson, R., J. Buurman, B. Newell, M. Boruah, C. C. Joon, A. Saikia, A. Bansal et al. 2017. *A system dynamics analysis of the levee effect on the Brahmaputra river and policy implications*. Proceedings of International Conference on Public Policy. Singapore: Lee Kuan Yew School of Public Policy.

Wester, P. 2003. Boundaries of consent: Stakeholder representation in river basin management in Mexico and South Africa. *World Development* 31(5):797–812.

Wirsing, R. G. 2008. *Rivers in contention: Is there a water war in South Asia's future?* Working Paper No. 41. Heidelberg Papers in South Asian and Comparative Politics. Heidelberg, Germany: South Asia Institute Department of Political Science, University of Heidelberg.

Zaman, M. 2012. Economic opportunities from polders in Bangladesh. *International Journal of Water Resources Development* 1(3):197–204.

Index

Note: Page numbers in italic and bold refer to figures and tables respectively.

action research 35
Actor-Network Theory (ANT) 53–4
adaptation strategy 110; action research 35–6; characteristics **36–7**; climate change governance 5–6; data analysis 37–8, **38**; deliberative governance 32; exploring deliberation 31–3; irrigation systems 34; methodology 34–8; organizations 36–7; overview 29–31; research area, Tarai-Madhesh plains 34–5; vulnerability, contrast framing 38–42
adaptive governance approach 93
Amazonian river 19
ANT (Actor-Network Theory) 53–4
Anthropology and Climate Change: From Encounters to Actions (book) 15
Assam Embankment and Drainage Act 169–70

Bagmati river 51–2
Bangladesh 71; agriculture 73; climate change and migration in 78–82; floods and people 72–4; governance 82–3; riverbank erosion 73
BCCSAP (Climate Change Strategy and Action Plan) 81
BEE (Bureau Energy Efficiency) 142
Bhaktapur Development Plan (BDP) 58
Bhaktapur, wastewater irrigation in 58; canal 58; farmers characterization **60**; Golmadi 61; rajkulo 58–9; sewage pipe 58–63
Brahmaputra 166–74
Brahmaputra River basin (BRB) 7, 165; barriers and bridges 174; debates 171; drainage and sub-basins *167*; knowledge gap 174–5; misfit river 166–74; opportunities 175–7; sustainability challenge 173–4
Bureau Energy Efficiency (BEE) 142

caste differentiation in education 122–6
climate change 3; adaptation policies 30; anthropology of 15; Ganga and human-water relationships 13–15; GOI on 11; migration and 74–7; social capital and adaptation to 92–3; Uttarkashi district 16–19
climate change governance 2; adaptation strategy *see* adaptation strategy; anthropology 15; challenges 3–5; competence/effectiveness 33; deliberative processes 31–3; fairness 33; migration, Bangladesh 71; national level and regional level 6–7; Nepal's political involvement 38–40; principles 33; prompt action 1; structure in India 139–40, *140*; vulnerability chain 40–2, *41*
Climate Change Policy in Nepal 38–40
Climate Change Strategy and Action Plan (BCCSAP) 81
climate refugees 71
communicative action theory 32
competence, deliberative governance 33

Dadhikot, wastewater irrigation 61; changing water sources 66; farmers characterization **62**; irrigation sources 63–4, *64*; kulo availability 62; rights and maintenance 64

deliberative governance 31–3; action research 35–6; data analysis 37–8; district level 43–5; limitations 38; national dialogue 45–6; public dialogues, organization 36–7; research area 34–5; VDC level 42–3
Dhanusa District *34*, 34–5
Dhara Vikas Yojana program 173
Disaster Risk Reduction (DRR) 164
Dolepchen GPU 121
DRR (Disaster Risk Reduction) 164
dzumsa 113n2

EACH-FOR (Environmental Change and Forced Migration Scenarios) project 76
Eastern Himalaya 110
ECA (Energy Conservation Act) 141, 143
education and adaptive capacity 111
Electric Rickshaws 149
Energy Conservation Act (ECA) 141, 143
Energy Saving Certificates (ESCerts) 142
Environmental Change and Forced Migration Scenarios (EACH-FOR) project 76
ESCerts (Energy Saving Certificates) 142
ethnography 3, 15

fairness, deliberative governance 33
farmer managed irrigation systems (FMIS) 96
farmers' films vulnerabilities 40–2
Focus Group Discussions (FGDs) 6, 114, **116**

Ganga *see* River Ganga
Ganges-Brahmaputra-Meghna delta *72*, *163*
Gangotri glacier 13
Garhwal 12, 16–17, 24
gender differentiation in education 126–30
German Association for Technical Cooperation (GTZ) 58
ghat 17
glacial melt 14
Global South, deliberative governance in 31–3
GOI (Government of India) 11
Goldilocks level 17
Golmadi 54–5, 61
GoN (Government of Nepal) 38–9
Government of India (GOI) 11
Government of Nepal (GoN) 38–9
gram panchayat units (GPU) 114, **115**; Dolepchen 121; of East District **115**; sub-components of education *122*; Sudunglakha 121–2
green initiatives in India 141; adaptation strategies and multilevel 153–5; in building sector 151–3; in industry sector 141–4; in power generation sector 145–7; in transport sector 147–51

hanga 100
Hindu beliefs in Ganga 17, 22–3, 90n1
Hindu Kush Himalaya (HKH) region 5, 90
HKH region, social capital in **95**; government interventions, growing 100–1; irrigation systems 96–7; male out-migration, increased 98–100; migratory pastoralism in rangelands 97–8; post-disaster settings 101–2; resources availability/productivity, decreased 95–8
human-river interactions 19–21
human-water relationships 13–15

ICIMOD (International Center for Integrated Mountain Development) literature 90, 93–4, 99, 101
IFAD (International Fund for Agricultural Development) 117
Igarapé Guariba 19–20
India: climate governance structure in 139–40, *140*; green initiatives in *see* green initiatives in India
Indian Himalaya: acts of worship, river Ganga 17; dwelling perspective 16; restate 16–17; river 16

Integrated Water Resource Management (IWRM) 165, 175
Integrated Watershed Management Program (IWMP) 153–4, *155*
Intergovernmental Panel on Climate Change (IPCC) 13
intermediate mode of transport (IMT) 147–8
International Center for Integrated Mountain Development (ICIMOD) literature 90, 93–4, 99, 101
International Fund for Agricultural Development (IFAD) 117
International Union for Conservation of Nature and Natural Resources (IUCN) 175
IPCC (Intergovernmental Panel on Climate Change) 13
irrigation systems 96–7
IUCN (International Union for Conservation of Nature and Natural Resources) 175
IWMP (Integrated Watershed Management Program) 153–4, *155*
IWRM (Integrated Water Resource Management) 165, 175

Kanchenjunga Conservation Area 98
Kathmandu Valley 51–2
Kathmandu Valley, wastewater irrigation in 56–8; ANT 53–4; Bagmati river 51–2; Bhaktapur 58–63; Bhaktapur municipality 54–5; Dadhikot 61–4; ethnographic approaches 54; kulo 55; peri-urbanity 52; political ecology 53; rajkulo 56; structural-relational mechanisms 66; Suryabinayak municipality 55
kulo 62–3, 66

labor out-migration 98–9
LAPAs (Local Adaptation Plans for Action) 40
least-developed countries (LDCs) 30
levee effect 169
Local Adaptation Plans for Action (LAPAs) 40

Mahatma Gandhi Rural Guarantee Scheme (MGNREGA) 173
male out-migration 98–100
mountain communities: post-disaster recovery of 101–2; service institutions 94
Multidimensional Poverty Assessment Tool (MPAT) framework 117
multi-utility system 145

NAPA (National Adaptation Program of Action) 30, 38–40, *39*
National Action Plan on Climate Change (NAPCC) 141
National Adaptation Program of Action (NAPA) 30, 38–40, *39*
National Ganges (Ganga) River Basin Authority 12
National Institute of Disaster Management (NIDM) 114
Nationally Determined Contributions (NDCs) 138
National Mission for Enhanced Energy Efficiency (NMEEE) 152
National River of India 12
National Rural Employment Guarantee Act (2005) 129n4
National Sample Census of Agriculture (NSCA) 34
NDCs (Nationally Determined Contributions) 138
NEMAF (Nepal Madhesh Foundation) 36
Nepal Forum of Environmental Journalists (NEFEJ) 35–6
Nepal Madhesh Foundation (NEMAF) 36
Nepal's Climate Change Policy 38–40, *39*
NERCORMP (North Eastern Region Community Resource Management Project) 101
NIDM (National Institute of Disaster Management) 114
NMEEE (National Mission for Enhanced Energy Efficiency) 152

North Eastern Region Community Resource Management Project (NERCORMP) 101
North-East Region (NER) 166
NSCA (National Sample Census of Agriculture) 34

Panchakanya irrigation system 96–7
panchanadi 96
Paris Agreement 138
participatory action research 35
Perform, achieve, and trade (PAT) 141–2, 144
peri-urban area 52; communities 4
pilot participatory video project 35
poly-generation system 145–6, *147*
post-empiricist approach 32
prakriti 23
public dialogues **36**, 36–7

Qualitative Comparative Analysis (QCA) 102n2

Raffles, H. 19
rajkulo 56–7
rangelands pastoralism 97–8
Right to Education Act (RTE) 120
river 16
River Ganga 12; climate change perceptions 19–21; cultural mediation 15; cyclical manifestation *18*, *19*; flooding in 24; glaciers of 13–14; goddess 17; Himalayan regions 14; Hindu beliefs in 17, 22–3; IPCC 13; monsoon rains 14; religious interpretations 22–3; symbol 22; in Uttarkashi *18*, *19*
RMDD (Rural Management Development Department) 173
RTE (Right to Education Act) 120
Rural Management Development Department (RMDD) 173

Samudayako Aawaj 36
Saraswati River 22
Sarva Shiksha Abhiyan (SSA) 120
SDAs (State Designated Agency) 143
SDGs (Sustainable Development Goals) 164
SEC (Specific Energy Consumption) 142
SECF (state Energy Conservation Fund) 143
Self Help Groups (SHGs) 94
Sendai Framework for Disaster Risk Reduction (DRR) 164
shakti 23
SHGs (Self Help Groups) 94
SIAS (South Asia Institute of Advanced Studies) 37
Sikkim 112–13
Sikkim, education status *112*, 120–2; agricultural productivity 113; caste differentiation in 122–6; children in agriculture *125*; data analysis 117–19; data collection methods/ techniques 114–16; Dolepchen GPU 121; dzumsa 113; gender differentiation in 126–30; gender inequality 113; household decision 129, *129*; literacy rate *124*; qualification across caste and gender *127*; study area/sites 112–14; Sudunglakha GPU 121
social capital 5–6, 90; and adaptation to climate change 92–3; adaptive governance 93; in community level application 93; in HKH region *see* HKH region, social capital in; methodology 93–5; women bonding 98–100
Solar Electric Rickshaw (Soleckshaw) 148
South Asia: climate change governance/ challenges 3–5; water in 161–5; water resources governance in 161–77
South Asia Institute of Advanced Studies (SIAS) 37
Specific Energy Consumption (SEC) 142
SSA (Sarva Shiksha Abhiyan) 120
State Designated Agency (SDAs) 143
state Energy Conservation Fund (SECF) 143

Subansiri dam 168
Sudunglakha GPU 121
Sustainable Development Goals (SDGs) 164

Tangkhul Naga 100
tripathaga 22

Ukhrul district 100
United Nations Framework Convention on Climate Change (UNFCCC) 30, 39
Uttarkashi District 12; climate change in 16–19; Ganga's flow in *18, 19*

village development committees (VDCs) 35, 42–3
vulnerability 110n1

wastewater irrigation: in Bhaktapur 58–61; in Dadhikot 61–4; in Kathmandu Valley 56–8; research approach, methods, and locations 53–6, *55*, **56**
water: culture 14–15; in South Asia 161–5
Water, Cultural Diversity, and Global Environmental Change (book) 14
Water Resources Act (1992) 66
women bonding social capital 98–100

For Product Safety Concerns and Information please contact our EU
representative GPSR@taylorandfrancis.com
Taylor & Francis Verlag GmbH, Kaufingerstraße 24, 80331 München, Germany

www.ingramcontent.com/pod-product-compliance
Ingram Content Group UK Ltd.
Pitfield, Milton Keynes, MK11 3LW, UK
UKHW011013180926
475701UK00004B/712

* 9 7 8 0 3 6 7 5 7 1 1 2 2 *